**WH**

C000270257

OPUS General Editors

Keith Thomas *Humanities*

J. S. Weiner *Sciences*

# D. F. Owen

# What is Ecology?

Second edition

Oxford New York Toronto Melbourne
OXFORD UNIVERSITY PRESS
1980

*Oxford University Press, Walton Street, Oxford* OX2 6DP

OXFORD LONDON GLASGOW
NEW YORK TORONTO MELBOURNE WELLINGTON
KUALA LUMPUR SINGAPORE JAKARTA HONG KONG TOKYO
DELHI BOMBAY CALCUTTA MADRAS KARACHI
NAIROBI DAR ES SALAAM CAPE TOWN

© *D. F. Owen 1980*

*First edition published as an Oxford University Press paperback
1974 and simultaneously in a hardback edition
Paperback reprinted 1976
Second edition in paperback and hardback 1980*

*All rights reserved. No part of this publication may be reproduced,
stored in a retrieval system, or transmitted, in any form or by any
means, electronic, mechanical, photocopying, recording, or otherwise,
without the prior permission of Oxford University Press*

*The paperback edition is sold subject to the condition that it shall not,
by way of trade or otherwise, be lent, re-sold, hired out, or otherwise
circulated without the publisher's prior consent in any form of binding
or cover other than that in which it is published and without a similar
condition including this condition being imposed on the subsequent
purchaser*

**British Library Cataloguing in Publication Data**
*Owen, Denis Frank
What is ecology? – 2nd ed.
1. Ecology
I. Title
574.5 QH541
ISBN 0–19–219155–1
ISBN 0–19–289140–5 Pbk*

*Set by Filmtype Services Limited, Scarborough
and printed in Great Britain by
Hazell Watson & Viney Ltd, Aylesbury, Bucks*

# About this book

Books about ecology tend to fall into two categories. There are those providing a background to ecology as part of a course in biology, and intended for school and university students. They are rather technical and assume some knowledge of the anatomy, physiology, and classification of plants and animals, together with considerable competence in mathematics, and they are on the whole unsuitable for people who are not seeking a formal qualification in biology. Then there are books oriented towards man's environmental dilemma, which aim to show that if population increase, economic growth, industrial pollution, and the destruction of natural environments continue at the present rate, we shall be facing catastrophe by the end of the present century. Such books have attracted a good deal of attention and have perhaps been partly responsible for making the quality of the environment a major political issue in the developed industrial nations. This book is somewhere in between the two.

Articles in newspapers and magazines and programmes on television have made ecology a familiar word to people who a few years ago would not have heard of it. Nevertheless, there is much misunderstanding as to what ecology is, and I hope that in this book I have been able to put across the essential scope of the subject and to place man in an ecological framework.

Ecology as a scientific study is concerned with the complex relationships between plants and animals and their surroundings; how they interact with one another, and how their numbers are limited by the resources of the world. Man is an animal, and as such can be viewed in an ecological context. But man is unusual in that he utilizes a far greater proportion of the world's resources than any other animal, and is moreover far more abundant than any other creature of remotely comparable size. Many of man's activities are destructive to nature and some have produced effects that cannot now be reversed. It therefore becomes essential that we examine ourselves from an ecological point of view and try to place ourselves in a framework that is part of nature; but this is not possible until we understand how the natural world operates. My aim, therefore, is to examine all of the important ecological principles, and to see whether we can learn something of our own prospects for the future in the light of what is known about other animals and plants.

This book is intended for people who having heard that there is such a thing as ecology now want to know more about it. I have tried to explain ecology without assuming prior knowledge of biology. To achieve this object I have taken almost all my examples of ecological principles from situations most of us have met and from familiar plants and animals. In deciding to approach the subject in this way I have necessarily omitted information about many less familiar plants and animals, but I do not think this is serious. I have concentrated more on animals than on plants because since we ourselves are animals it is easier to understand them; and I have for the same reason picked on vertebrate animals rather than insects and other invertebrates wherever possible.

I have heard professional ecologists – those working in universities and research institutes – complain that the meaning of ecology has become blurred as a result of public interest in the subject. I do not share this view. There are of course dangers in writing about a highly technical subject in a way which should be comprehensible to all, but I consider the risk worth taking. I have in mind the interests of gardeners, anglers, farmers, students, and people in all walks of life who have heard about the dangers of over-population, pesticides, and the destruction of the countryside, and who simply need more background to ecology as a subject.

This new edition of a book first published in 1974, at a time when the world was counting the cost of the 1973 oil crisis, is almost completely re-written and expanded, to the extent that it is effectively a new book. I have in particular included new material on energy flow, plant/animal relationships, weather and climate, animal migration, desertification, and food production. Many of the problems of human ecology discussed in 1974 are still topical; some have been partly resolved; others are more acute; but there are encouraging signs of increased interest in the quality of the environment and there are one or two success stories.

I am grateful to Messrs. Longman, London for permission to

reproduce in Chapter 5 some of my own material from *Ecological Biology 1. Organisms and their environments* (edited by D. W. Ewer and J. B. Hall, 1972). I thank Richard Wiegert for a critical reading of an earlier draft and Jennifer Owen for much help with both the first and the revised editions. Richard Owen provided information about angling and Roger Whiteway supplied much of the material on buddleia. I thank also many friends and reviewers of the first edition who have supplied suggestions and ideas, some of which are incorporated in the present edition.

Denis Owen, 1979

# Acknowledgements

Figs. 1, 2, 3, and 4 Maps provided by J. Heath, Monks Wood Experimental Station; Figs. 6 and 10 Reproduced with permission from D. Lack (1954) *The natural regulation of animal numbers*, Clarendon Press, Oxford; Fig. 11 Reproduced with permission from D. Lack (1966) *Population studies of birds*, Clarendon Press, Oxford; Fig. 20 Reproduced with permission from R. H. Whittaker (1970) *Communities and ecosystems*, Macmillan; Fig. 21 Reproduced with permission from F. C. Evans and U. N. Lanham (1960) *Science* **131**, 1531; Fig. 23 Reproduced with permission from R. G. Wiegert and D. F. Owen, *J. Theoret. Biol.* **30**, 74 (1971); Fig. 27 Reproduced with permission from E. P. Odum, *Fundamentals of ecology*, Saunders 1971. Figs. 14 and 28 Reproduced with permission from The Zoological Society of London. Plates 1, 11, 12, and 16 Jennifer Owen; Plate 2 East Africa Common Services; Plate 3 John Hillelson Agency; Plate 4 Michael W. Richards/R.S.P.B.; Plate 5 Malcolm Coe; Plate 6 Graphic Photo Union; Plate 7 Sven-Axel Bengtson; Plate 8 Centre for Overseas Pest Research, London; Plate 9 Aerofilms Ltd; Plate 10 Arne Schiøtz; Plate 14 D. O. Chanter; Plate 15 Jennifer Owen and Keith Arnold; Plate 17 O. Hedberg; Plate 18 Shell International, London; Plate 19 W. R. Whiteway; Plate 20 H. B. D. Kettlewell; Plate 21 Charlotte Ward-Perkins; Plate 22 Bureau of Mines, U.S. Department of the Interior, and Billy David and the Louisville Courier-Journal; Plate 23 Politiken's Press, Copenhagen; Plate 24 R. K. Murton; Plate 25 Bruce Colman Ltd.

# Contents

# List of plates

# 1 What is ecology?

Ecology is concerned with the relationships between plants and animals and the environment in which they live. This simple explanation is the kind of answer a school child would offer if asked 'What is ecology?' But the explanation, although apparently neat and simple, does not specify what is meant by relationships and what is meant by environment. These two words occur throughout this book and we shall therefore begin by considering what they mean.

There are many possible kinds of relationships between organisms (plants, animals, and other living things like viruses) and that part of the non-living world in which they occur. An extremely important one is who eats whom, and another, perhaps equally important, is who breeds with whom.

The concept of the environment covers just about everything associated with organisms, and includes other organisms and the non-living part of the world in which life occurs. The weather, the physical and chemical composition of the soil, and seasonal changes in the length of daylight, are all parts of an organism's environment, and the word therefore has about the same meaning as surroundings.

No organism exists without an environment; organisms and the environments in which they live constitute an extremely thin layer on the surface of the earth, often called the biosphere, in which the very complexity of ecological relationships tends to frustrate scientific analysis. Nevertheless careful consideration reveals some order in the biosphere which can be understood and defined, although it must be admitted that we are still a long way from formulating general theories of the kind familiar to students of chemistry, physics, and mathematics.

The essential feature of living organisms as opposed to non-living objects is that they reproduce and replicate themselves. Organisms are associated with land or with water; many kinds spend the greater part of their life in the air, but none exists entirely in the air. The most familiar organisms, trees and plants, and the animals that feed on them, are obviously associated with the ground, but everyone knows that both fresh and salt water support a variety of plant and animal life.

The Greeks used the word *oikos* to describe a home, a place to which you could return and where you understood and were familar with the local environment. From this word we have derived the terms ecology and economics to describe the subjects concerned with aspects of home life. Ecology began as descriptive natural history but nowadays scientists study and describe ecological phenomena in quantitative terms, often to such an extent that scientific magazines devoted to ecology publish articles that look more like pages from textbooks in mathematics. Much the same applies to the scientific study of economics. But economics and ecology are subjects in which intuition also plays an important role, and although both subjects have received rigorous mathematical treatment it could be doubted if this has told us a great deal that we did not know already.

In recent years ecology has become a household word. It has begun to enter into discussions about economic development, industrial growth, and standards of living, but there is often confusion. Many people think ecology is another word for pollution or the conservation of rare animals; others see it as something of a political plot against economic growth. Only now have a substantial number of people become concerned about trends in the growth of the human population and the consumption of natural resources, and with this concern fears for the future are increasingly expressed. What exactly has generated this interest, and why has ecology, until recently a rather obscure subject, come to the forefront?

In industrial countries the most important single event has been the realization and demonstration that pollution resulting from industry and agriculture is harmful to people and to the surroundings in which people live. Industrial pollution comes from waste products of the manufacturing and service industries, and these products are difficult, or at least expensive, to dispose of, while agricultural pollution results from the accumulation of toxic chemicals derived from pesticides and fertilizers. But although the present awareness of ecology may have been stimulated by the apparent dangers of pollution, this is not primarily what ecology is about. It is a much bigger and more complex subject than this, and to understand ecology we must move away from this restricted though common viewpoint. Ecology is not even

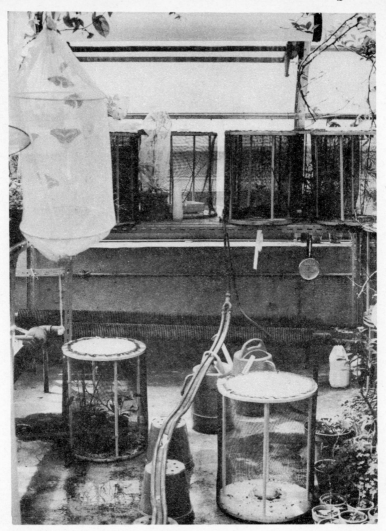

*Plate 1:* Inside an ecological laboratory. Tropical swallowtail butterflies are being bred under controlled temperatures and humidity in order to investigate the environmental factors that determine whether the pupae are green or brown. University of Leicester, 1972.

primarily about man; indeed it covers all living organisms, and although most people will undoubtedly be interested in the implications of ecology for human life and welfare, to understand the subject it is necessary to abandon an entirely man-oriented approach and to consider instead the inter-relationships of all life and the environment. Hopefully this will lead to a better understanding and more respect for plant and animal life and perhaps help to explain how the present lack of balance between man and the environment has come about.

Evidently, then, there are two facets to our subject. One is mainly but not entirely disinterested scientific inquiry, the other more diffuse and concerned with political and economic problems of over-population, the consumption of resources, pollution, conservation, and the plight of under-developed countries. Ecology as a science involves detailed and painstaking measurements of population sizes of plants and animals, birth and death rates, the supply and utilization of energy and nutrients in the environment, and related subjects. It is really a sophisticated and academic form of natural history. The other facet is more concerned with man's place in nature and threats to the quality of life. The two facets are compatible although curiously enough people active and vocal in one are rarely active and vocal in the other. Most scientific ecologists take no public stand on the wider, less clearly defined issues that afflict mankind, but apart from journalists and writers there are few professional 'political' ecologists.

One way to begin an understanding of a new subject is to inquire into the activities of those who practise it, and this is what we shall now attempt.

## What is an ecologist?

A satirical magazine recently defined an ecologist as a man with a beard who appears on television. Such a man is probably talking about human population trends, the future supply of natural resources, the 'energy crisis', or the conservation of interesting places and rare plants and animals. His viewpoint is almost certainly that of a pessimist, but before dismissing his arguments it is important to appreciate that pessimists are not necessarily always wrong any more than optimists are always right. Our television ecologist seems to speak with authority and specialized knowledge, and he is probably either a writer or a broadcaster on environmental issues, or, possibly, a scientist employed in a university or a government research organization. If he is a scientist his normal job is to teach others about ecology, or to indulge in ecological research, or both.

Put a group of professional ecologists together for a discussion and however much they argue they will end up by agreeing on one issue: every ecological investigation raises as many new questions as it answers old ones; indeed it appears that there are no limits to the possible lines of ecological research. A clever ecologist can, by asking pertinent questions, initiate dozens of lines of research which will keep lesser ecologists occupied for years. Thus in the nineteenth century A. R. Wallace observed that there are many more species* of plants and animals in the tropics than there are in the temperate regions and that many tropical species are relatively rare. There may be over a hundred species of trees and three hundred species of butterflies in a small patch of tropical forest, while in a temperate area of comparable size there will be perhaps twenty species of trees and only about thirty of butterflies. Many of the temperate species are common, but most of the tropical species are rare, so that total numbers of trees and butterflies may not be so very different in the two environments, but the diversity is quite different. Why should this be so? Questions like this have been raised repeatedly and ecologists have initiated much research and proposed several theories to explain the phenomenon, but its significance remains elusive.

## Natural history and ecology

It is certainly wrong to assume that the only ecologists are those professionally qualified and employed in universities and research organizations, or those who are vocal about the coming environmental crisis. We have already mentioned the relationship between ecology and natural history. Paricularly in Europe and North America, and especially in Britain, there are thousands of people in all walks of life who are passionately interested in natural history. Numerous natural history societies and clubs flourish. Amateur naturalists usually develop an interest in a particular group of plants or animals, and some groups, such as birds, butterflies, and wild flowers, are understandably more popular than others, but even the 'difficult' groups like earthworms and mosses, have their amateur specialists.

Amateur interest is less frequently oriented towards ecology as a technical or scientific subject. Thus although there are many bird-

* The species concept is explained in Chapter 4.

*Plate 2:* Ecology in the field: a tall steel tower built in the forest in Uganda by the East African Virus Research Institute. The tower is used by ecologists to study swarming and biting behaviour in mosquitoes and other flies that may transmit viruses.

watchers there are probably few amateurs interested in such topics as the energy relationships in environments, not because these topics are necessarily difficult to understand, but because for detailed study they require rather specialized equipment and facilities. In addition it seems that much amateur interest satisfies the urge to collect, whether the names of birds seen or specimens of beetles. Few amateurs embark on an investigation in order to answer a question: their work is mainly descriptive and observational.

In countries like Britain much of what is now known of the distribution of familiar plants and animals, and of changes in distribution, has been obtained through the efforts of amateur naturalists. Some societies like the London Natural History Society have for years been conducting detailed surveys of specific areas, so that more is probably known about London's ecology than that of any other place in the world.

The study and collecting of butterflies has been a popular activity for more than two hundred years, and the life histories, distribution, and behaviour of British butterflies are well known. Nowadays with the disappearance of many butterflies from areas where they used to flourish there is a welcome emphasis on conservation rather than on collecting. Largely as a result of the efforts of amateurs, professional ecologists at Monks Wood Experimental Station (a government centre devoted to ecological research and conservation) have been able to map in detail the present and past distribution of all the species of British butterflies. Fig. 1 is a map showing the distribution of the brimstone butterfly, a distinctive species with a lemon-yellow male and a greenish-white female. In the map the whole of Britain is divided up into 10 km squares and a black dot means that brimstones have been recorded in that square since 1960 while a circle means that they were recorded before 1960. The absence of a dot does not necessarily mean that the butterfly is absent from the area; it might mean that no one has yet recorded it there, and in Fig. 1 the missing dots in areas like south-east England are no doubt because of lack of records. But the absence of dots and circles in most of Scotland, northern England, and Wales, and the few on the east coast north of The Wash indicate absence or rarity.

The brimstone lays its eggs on two kinds of shrub, the buckthorn and the alder buckthorn, and the caterpillars feed on the leaves of these shrubs. Like many butterflies, brimstones are highly selective, and in

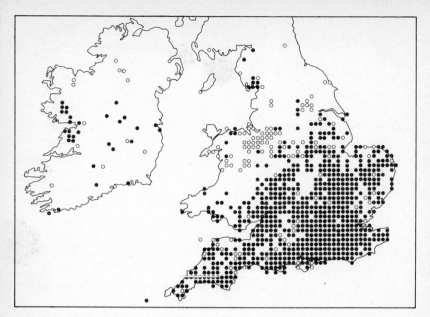

*Fig. 1:* The distribution of the brimstone butterfly in Britain. Black dots refer to records from 1960 onwards and circles to earlier records.

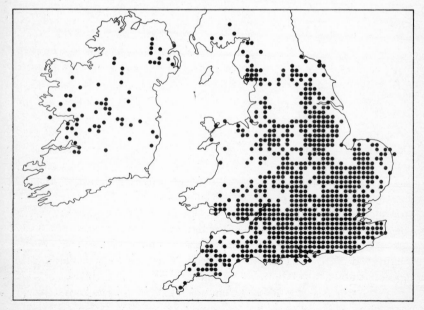

*Fig. 2:* The distribution of buckthorn and alder buckthorn, larval food-plants of the brimstone butterfly, in Britain.

Britain they feed only rarely on other species of plants. Fig. 2 shows the distribution of the two food-plants in Britain, and it is immediately apparent that the distribution of the butterfly and its food-plants is similar and strongly correlated. Ignoring the unlikely possibility that the distribution of the butterfly determines the distribution of the food-plants, there remain two possible explanations for the correlation. The first and most likely is that the butterfly is found only where its food-plant grows and is unable to survive where it is absent. But it is also possible that a third factor, perhaps climatic, is determining the distribution of the butterfly and the plants, which in effect is saying that the occurrence of the plants does not affect the butterfly, but that both are independently affected by an unknown third variable. Correlations of this sort often occur in ecology and they are notoriously difficult to interpret. The final word cannot be given, but what is important here is that basic information obtained by amateur naturalists was brought together by professional ecologists, a good example of collaboration between people with different skills and interests.

The distribution of many other species of insects has been mapped as a result of such collaboration. Some of the distributions are difficult to interpret: for example, the black hairstreak butterfly occurs in woods between Oxford and Peterborough in the area shown in Fig. 3; occasional records outside this area are the result of deliberate introductions or of mistaken identity. The caterpillars feed on the leaves of blackthorn growing in sunny woodland glades and in old hedges. Blackthorn is one of the commonest shrubs in Britain and occurs almost everywhere, as shown in Fig. 4. All over the country there are suitable-looking places yet no black hairstreaks. Thus unlike the case of the brimstone there is no correlation between the distribution of the butterfly and its food-plant. We must therefore seek an alternative explanation to account for the distribution of this species. Various suggestions have been put forward. Is there something peculiar about the climate of the area, or the soil or geology? Is there a historical explanation – perhaps the butterfly was once more widely distributed, and as the woodlands were cut down by man became more and more restricted and isolated? Perhaps there is something peculiar about blackthorn which makes it palatable to black hair-streak caterpillars in some areas but not in others? No one knows the answer to these questions; indeed we do not even know if they are the

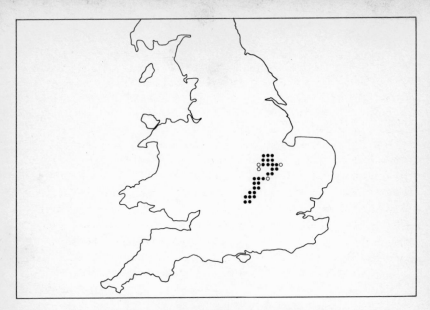

*Fig. 3:* The distribution of the black hairstreak butterfly in Britain. Black dots refer to records from 1960 onwards and circles to earlier records.

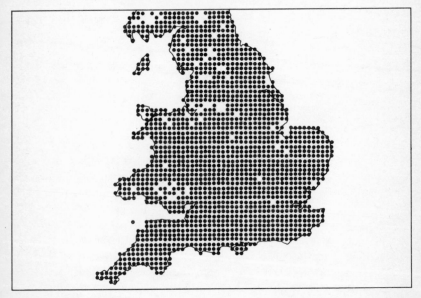

*Fig. 4:* The distribution of blackthorn, the larval food-plant of the black hairstreak butterfly, in Britain.

right questions, but we can be sure that there is an explanation for the strange distribution of the black hairstreak in England, just as there must be for the distribution of all species of plants and animals in all parts of the world.

## Everyone is an ecologist

Man is a large and abundant animal living in almost every part of the world. Before the invention of agriculture and the domestication of animals like cows man was a hunter and a gatherer. It is estimated that a hunter needs about twenty square kilometres to sustain himself and yet the same area under cultivation can support about six thousand people. This figure need not be taken as exact but it serves to show how the invention of agriculture has made possible large human populations.

The development of world agriculture and the consequent rise in the human population has been brought about by what has been (and still is) essentially a trial-and-error ecological process. Crops are constantly tried in new environments, and those that survive and yield well are preserved and perpetuated for future use. A prime consideration for peasant cultivators and highly mechanized farmers alike is that of finding the most productive crop for particular local conditions. Today ecological knowledge acquired over centuries is used by the farmer to decide which crops will do best in a particular field, and on a smaller scale by the gardener planning flowers and vegetables for his garden.

Some areas of the world can produce high yields of crops, especially with the application of artificial nutrients in the form of fertilizers, and pesticides to control the numbers of harmful animals, but other areas of the world are less fertile, and agricultural activities can lead to barren land which so far as man is concerned may become almost totally unproductive.

Agriculture began in Asia about ten thousand years ago and quickly spread westwards and eventually southwards into Africa. It may have originated independently in South America or it may have spread there from Asia via North America. Agricultural crops are derived from wild species of plants but in many instances the wild ancestor of our common crops is not known. Not all crops can be grown on a particular piece of land; some survive only in tropical environments, some require a cold winter, some an acid soil, and so on, and the present distribution and diversity of crops in different regions of the world is the result of repeated trials with many failures and some successes. This process is continuing, but with increased scientific knowledge it has become easier to predict which crops are likely to succeed in a particular environment and which are likely to give the best yields.

A modern farmer has available a vast store of information about which crops are best grown on his land. He will also be able to make judgements about the prospects of selling his produce at a good price and in response to demand. Even so environmental conditions vary; no two years are exactly alike in terms of climate, the condition of the soil, and losses through disease and pests. His interest is to get the best yield with the minimum of expenditure and the maximum of profit, and in deciding what to grow he will take into consideration as many factors as possible, some of them economic, but many of them ecological.

A peasant cultivator in the tropics has by necessity to be even more of an ecologist. He lives in an environment where traditions have been passed down by word of mouth generation after generation, and where everyone is poor. He grows just enough for the family to live on, and lacks exact technical knowledge of what is best to grow. But he will be able to judge accurately from the natural vegetation which piece of land is worth clearing and cultivating. He will have acquired a vocabulary of hundreds of names for plants, trees, and vegetation associations, and will be fully aware of the significance of indicator plants that tell him about the moisture-holding properties of the soil and which crops are worth trying. His knowledge of the local climate, particularly as regards the seasonal cycle of rain and drought, will be indispensible. None of this knowledge is written down (indeed even if it were he would probably be unable to read it) and everything he knows he has learned from others, especially from his own family or village, or from his own hard-won experience. Scientific ecologists would be delighted at his powers of prediction, and in every sense he is an expert. But if displaced from his own environment he will face disaster, because his knowledge is no longer of value. Everything he knows has been acquired through pressures exerted by his surroundings and without his ecological knowledge his own position and that of his family would at best be precarious.

A feature of peasant cultivation is the apparently haphazard arrangement of different sorts of crop. Maize, onions, sweet potatoes, melons, tomatoes, and so on may be grown intermixed. An agricultural expert from a temperate country sees this arrangement as untidy, inefficient, and unproductive, and may advocate plots of a single crop laid out in neat rows. However, medieval European gardens incorporated mixed planting, and at the Chateau de Villandry in France the centuries-old gardens still have vegetables mixed with herbs and flowers. The only legacy city-dwellers have of mixed planting is the traditional herbaceous border with its riot of colours and scents. Neat, ordered monocultures are easier to harvest and treat with pesticides and so are favoured by gardeners and farmers in temperate countries. But is it possible that the problems created by pests and weeds are the result of our modern methods of cultivation?

Increasingly credence is given to the idea of companion plants. Evidence is accumulating that some plants deter insect pests and disease-causing organisms, others inhibit weed growth, and many concentrate essential nutrients which become available to crops when the plants die and decompose on or near the surface of the soil. Work on hemp in Russia suggests that it produces chemical compounds that deter mole crickets, which are troublesome pests as they eat the roots of maize, tomatoes, and potatoes. If hemp is interplanted with crops the pests disappear. Cabbages in the United States and elsewhere are attacked by tiny flea-beetles; experiments show that these beetles locate the cabbage plants by smell, but are less able to do so if aromatic plants like tomato are grown nearby. It appears that an insect seeking food or egg-laying sites by smell is confused by a battery of conflicting aromas. Some strains of cucumber give off from their roots an unidentified chemical substance which severely restricts weed growth, even when washed out of the soil around the cucumber plants and applied as a spray. Experiments such as these confirm the belief in the value of companion plants. Modern gardeners are beginning to learn of the value of growing comfrey for compost. The plants concentrate potassium from the soil and accumulate protein faster than other species; some enthusiasts see comfrey as the solution to the problem of buying and using artificial fertilizers.

Indeed it is possible for the gardener, however tiny his plot, to arrange his garden in such a way that he incorporates knowledge of plant and animal relationships with a great deal of ecological precision. The possibilities are endless: planting near aphid-infested rose bushes flowers like marigolds that attract hoverflies has much to recommend it, as the larvae of hoverflies are voracious predators of aphids. Leaving weedy patches near vegetables will provide shelter for beetles and spiders which at night may prey on the pests of the vegetables. But not all of the companion plant thesis makes such good sense. Some advocates believe that plants 'help' one another in unspecified ways while others think that the scents from herbs invigorate the atmosphere. As with all beliefs we are probably dealing with a mixture of fact and fallacy, and there is something of a challenge to separate the two. But is it not likely that the peasant cultivator in the tropics, with little or no access to pesticides, herbicides, and artificial fertilizers, has by trial and error worked out the value of certain plant associations? He is after all interested in a sustainable yield to feed his family and cannot afford the risk of losing the entire crop, which might occur if he adopted a more tidy approach.

In one way or another we all use the kind of ecological knowledge which to the peasant cultivator is so important for survival, but in most cases this knowledge is not important for our own survival. Many of us use ecological knowledge during outdoor leisure activities, especially in such hobbies as gardening and angling, both popular among people who earn their living in factories and offices well insulated from the natural world, and possibly pursued because of a fundamental need to maintain contact with nature.

## Looking after a lawn

Grass lawns have been in existence in Europe for centuries. They are decorative and aesthetically pleasing but they are in no sense important for survival; nor indeed are they economically important in more than a trivial way. They may possibly have been of considerable social significance in the past, and even today the village green and the private lawn are places where people meet and talk in the open away from concealing bushes and trees. Europeans have transported the idea of creating lawns to almost all parts of the world. In Africa, for example, Europeans have established lawns around their houses and in public places and the tradition has been copied by affluent Africans. In November 1976, following a dry spell, Maasai herdsmen drove their cattle into Nairobi and grazed them on the sprinkler-fed grasslands surrounding government buildings – until the police intervened.

A lawn is a patch of grass which is constantly cut short. There may be several species of grass and some kinds make better lawns than others. A lawn in the United States may be sown with the same grasses as an English lawn, but different species are used in the tropics. Lawns flourish best in cool climates where the rainfall is light and persistent, which explains why English lawns are much admired by foreign visitors. A good lawn does not contain plants other than grass, but the creation of a lawn provides an environment which favours colonization by other plants commonly regarded as weeds. A weed is a plant whose presence is unwelcome either because it inhibits agriculture or horticulture or because it is aesthetically unpleasing. Different people in different circumstances will therefore have varying views as to what constitutes a weed: a cabbage is not a weed in a cabbage-patch, but if it grew in the middle of a lawn it would be removed. Lawns are maintained by cutting and rolling the grass at regular intervals, removing weeds like plantains and dandelions by digging them out or by the application of herbicides, re-seeding or re-planting patches where the grass has disappeared, and watering whenever there is not enough rain. If these and other tasks are not performed the essential character of the lawn quickly disappears.

The growth of grass in a temperate climate follows a seasonal pattern, and in winter there is normally no need to mow the lawn or to remove weeds. Growth is most rapid in summer and when the weather is wet. Mowing the grass with a lawn-mower prevents it from flowering and seeding and therefore encourages the vegetative spread of grass. Most people use lawn mowers that collect the grass cuttings. The cuttings are then removed, often to a compost heap where the dead grass decomposes, and later on may be spread on ground used to grow flowers or vegetables. In ecological terms this means that a substantial proportion of the living material produced by the lawn is being removed and there is a steady loss of nutrients from the environment. Nearly all that a lawn produces is ultimately removed in the form of cut grass or litter and it may eventually be necessary to apply artificial fertilizers. These may not replace all the nitrogen, phosphorus, and potassium that have been removed, but since artificial fertilizers contain these substances in a form that can quickly be taken up by grass roots, they correct any loss of 'quality' of the grass. Gardeners, it may be added, have an intuitive feeling for grass quality, although they would find it difficult to define and explain.

During the cutting of the grass any weeds present are also cut, and some like dandelions respond to repeated cutting by flowering on a short stem, far shorter than they would if there were no cutting. In a sense, then, the lawn-mower maintains a form of selection which favours individuals and species that can produce flowers and seeds close to ground level, and this is why the small daisies characteristic of English lawns are so successful. If weeds are not removed they spread and become conspicuous in the lawn, but repeated cutting of the grass and removal of weeds results in the establishment of an area of land dominated by a few species, and any natural tendency of the area to support more species is inhibited by the gardener, provided, of course, he is doing his job well. If for some reason lawn-mowings are not removed quantities of decomposable plant material are left lying around and favour establishment of populations of animals like slugs and woodlice. The decomposition of the mowings proceeds at a rate dependent on the dampness of the lawn and the prevailing tempera-ture. Watering a lawn creates an environment favourable to earth-worms, which are also decomposers of dead vegetation, and most lawns support large populations of these animals. Because the grass is kept short earthworms are available to birds that feed in rather open situations and are able to exploit food sources on or just below the ground. In Europe one of the most conspicuous of these is the song thrush, an efficient earthworm feeder, and gardens with lawns provide excellent breeding places for these birds, as well as for other worm-eaters like blackbirds.

An ecologist in California monitored for a year the work and cost needed to maintain a small suburban lawn, and measured the production of vegetation and the abundance and variety of animals dependent on it. He established some interesting differences between a lawn and natural grassland: in a lawn the quantity of live grass exceeds that of dead grass whereas in uncut natural grassland the converse is true. However, during the year the gardener removes as cut grass the bulk of the vegetation produced, and is very much the dominant animal in terms of impact on the lawn and its associated flora and fauna. In contrast to lawns, grassland may show decreased pro-ductivity if it is cut. One way in which ecologists measure the productivity of a habitat is by weighing the vegetation produced in a year. The value for the Californian lawn slightly exceeds the value for natural grassland and compares favourably with the value for fields of maize.

The great variety of small invertebrate animals in the Californian lawn falls into two categories: those small and mobile enough to sit on grass blades and escape when disturbed, and those able to seek refuge in the litter hidden from view and cushioned from trampling. Despite mowing, trampling, and the application of pesticides, the efficiency with which the plant-feeding animals of the lawn utilize the vegetation is about the same as for natural grassland. But there are fewer predators on the lawn, particularly spiders, and the efficiency with which predators remove the plant-feeders is low. Evidently lawn management, including pesticide application, destroys more pre-dators than the plant-feeders that the gardener may consider as pests.

The Californian study also confirms that lawns are enormously important as feeding sites for birds; they can extract several times the quantity of food that they can get from an equivalent area of natural grassland. The explanation for this lies in the very nature of lawns: birds can feed on every square centimetre, with an unobstructed view of impending danger, which they cannot do easily in natural grassland.

Ecologists are increasingly interested in energetics, in other words, energy inputs, pathways, and outputs. Energy is usually expressed in units of heating capability (e.g. joules or calories), but in real life may take many forms – labour, petrol, power used to manufacture pesticides or pump water, and so on. The major source of energy input for the Californian lawn is in watering, which may not be necessary in wetter, cooler localities. The gardener's total energy input in main-taining a square metre of lawn is greater than that required to cultivate a square metre of maize. Indeed, a lawn uses as much energy as a vegetable patch and as energy and vegetables become more expensive the tradition of keeping lawns may decline. In Britain in 1974, according to a recent Government report, there were nearly 13 million domestic lawns covering 90 thousand hectares and costing more than £20 million to maintain. Can we continue to afford this luxury?

The kinds of grasses useful for the establishment of a lawn and the kinds of weeds and animals which invade a lawn will of course depend upon the locality. It is notoriously more difficult to start to cultivate a lawn in the tropics than in the temperate regions, and this is probably because there is a much greater diversity of species of plants and animals which are potential colonizers of the lawn. Earthworms are rather scarce in the tropics and rarely colonize tropical lawns. Their place as decomposers of dead vegetation is taken by termites, a group of insects superficially similar to ants.

Each gardener has his own ideas as to what his lawn should look like and each takes whatever action seems appropriate to maintain his concept of a lawn. Thus he may use mercury compounds to rid the lawn of moss if he considers moss undesirable, ignoring the possible toxic effects of mercury on animals that he might otherwise enjoy seeing in the garden. Lawns do not under normal circumstances maintain themselves; if left unattended they quickly revert to a quite different association of plants, and so the gardener, using his knowledge of local conditions and keeping in mind what he wants, has to give the lawn constant attention. A Canadian government scientist was so sympathetic with the vain efforts of residents in Arctic Canada to maintain traditional lawns when faced with snow, cold, and salt used to de-ice paths, that he devised a rapid repair technique for damaged lawns: replacement sods based on cheesecloth are cultivated indoors then fitted into the lawn as appropriate.

In maintaining a lawn the gardener is quick to seize upon new ideas and technological innovations: he also makes use of knowledge of the relationships between organisms and the environment, and we should therefore have little hesitation in calling him an ecologist, although he might not think of himself in these terms.

## Catching a fish

Angling is a popular hobby and thousands of people from all walks of life set out at the weekends to try their luck. In Britain a few of the fish caught, like salmon and trout, are kept for food, but most anglers return to the water many of the fish they catch. One group called 'coarse' fish is especially popular with British anglers. It includes familiar species like roach, perch, tench, carp, bream, and pike. None of these species is normally eaten in Britain, partly through prejudice and partly because they are apt to taste muddy; with the exception of pike they are returned alive to the waters from which they were taken. Many anglers would regard it as morally wrong not to return fish like roach and would argue that it is only by returning fish that the populations are conserved for future sport. The pike occupies a rather peculiar position in the eyes of British anglers. It is a predator of other coarse fish and is therefore regarded as something of an enemy of the angler, and although it is sometimes kept for eating, it is often killed and thrown away. In North America, where the same species of pike occurs, and elsewhere in Europe, it is regarded as perfectly good human food.

In Britain there is a close season on coarse fishing from 14 March to 16 June, the breeding season of most species of fish. The idea is that the fish and their breeding areas are left undisturbed, but there are at present discussions in the angling newspapers as to whether the close season is necessary or desirable from the point of view of conserving fish stocks. Each time a fish breeds it produces thousands of eggs, up to three hundred thousand in the case of the perch, and half a million in the case of a large pike. It follows that since the waters do not become choked with fish almost all the eggs fail to produce even small fish. There must be an enormous mortality soon after the breeding season, possibly brought about by a shortage of food or by the activities of predators, including large fish and some kinds of aquatic insects. In view of this mortality it could be seriously doubted whether it is necessary to maintain the close season. The numbers of fish removed by anglers are presumably small relative to the number of fish in the water, and minute compared to the number of eggs produced, and the main justification for the close season must be seen as an attempt to leave the breeding grounds undisturbed.

Returning live fish is also of dubious value, and the fact that this is common practice does not necessarily mean that it is sound from the point of view of conservation. Fish, like some other groups of animals, will, if crowded, become stunted and breed when small, and in essence the angler has the choice of more small fish or fewer large ones. Removal of fish should result in those that remain growing larger which many anglers would consider desirable, as large fish are at a premium and are much sought after for sport. Every week the angling newspapers are full of photographs, and it is always the larger fish that receive publicity. For some stretches of water containing high densities of coarse fish there is little doubt that the average size could be increased by reducing the number of fish present. Furthermore many fish are damaged during catching. The hook itself can inflict injury to the mouth, no matter how carefully it is removed, and scales are inevitably lost during handling. This damage could reduce the chances of the fish surviving once it has been returned, for it frequently provides a site for infection with fungus diseases. A diseased and damaged fish is not pleasant to catch and moreover it is much more likely to be taken by predators. A fish that has been lying in a keep net for hours waiting to be weighed and released will behave abnormally when introduced back into the water and may as a result fall an easy victim to a predator simply because of its unusual behaviour.

Freshwater fish are an important sporting resource and it is therefore desirable that stocks are maintained at an optimum level and condition. It seems likely that both the close season and the custom of returning fish to the water are unsound policies from the standpoint of conservation, but there is considerable need for study here, and the best policy for one stretch of water may not necessarily be the best for another.

There is a striking contrast between the application of ecological knowledge in catching fish and in maintaining lawns. An experienced gardener has precise knowledge of the best way to keep his lawn in good condition, but the angler is much more uncertain about the waters in which he fishes, even though he may claim considerable knowledge. This is presumably because fish live in water and their activities and needs are less easily seen, and so the angler's attitude is based more on belief and guess-work than on knowledge. There is probably no other outdoor activity that involves as much hearsay and belief as angling. But anglers, who constitute a powerful conservation lobby equalled only by bird-watchers, are much quicker to detect and complain about the effects of pollution than gardeners, some of whom are inexplicably careless in the use of pesticides and fertilizers. If a river or lake becomes polluted, anglers soon detect what has happened, and they frequently find themselves able to supply advance information to people whose job it is to assess the level of pollution in waters. In other words familiarity with local fishing conditions is often a better source of information about the water than even the most elaborate chemical tests.

Both the gardener and the angler adopt an ecological approach to their respective environments, the gardener judging the lawn by repeated examination and appraisal of what he considers desirable, the angler judging the water in terms of the size and species-composition of his catch, and both in their own ways are in a position to modify the environments in which they have interests.

We have been speaking of leisure activities in which participation is solely for enjoyment, and obviously neither gardener nor angler depends on his special interest for survival. Their personal resources probably come from their jobs which provide not only enough to support themselves and their families, but also for them to indulge in their hobbies. In the next section we examine the ecological relationships between two kinds of animal and the environment in which they live. We shall be much closer in this example to questions of survival, although we shall necessarily restrict the discussion to only part of the story. As will become evident later, ecological relationships are intricate and varied and we shall therefore begin with some simplifications which regrettably will result in an element of distortion.

## Thrushes and earthworms

As everyone knows earthworms are burrowing animals living near the surface of the soil and coming to the surface chiefly in wet weather and at night. They occur in most parts of the world and in temperate grassland may be exceedingly abundant, with a hundred or more to a square metre. All earthworms look alike, but there are in fact many species, and an average garden contains quite an assortment, although most gardeners are unlikely to be able to tell them apart. Earthworms feed on dead and rotting vegetation and by means of a muscular pharynx suck up large quantities of soil containing humus, pass it through the gut, and leave distinctive castings scattered over the surface of grassland. They are so abundant and they pass so much soil through their bodies while feeding that their activities bury hard objects like stones, and this is one reason why archaeologists have to dig for evidence of ancient buildings.

Many animals eat earthworms, but it is perhaps remarkable that only a few human societies incorporate them into the diet despite the fact that they are a good source of protein. The most efficient earthworm-eaters are probably animals that habitually burrow in the soil, like certain snakes, legless lizards, and moles, but other non-burrowing animals eat them when they reveal their presence on or near the surface of the ground. Among these are several species of birds, the most conspicuous and efficient of these in Europe being the song thrush. This bird has a remarkable ability for detecting even the tip of an earthworm sticking up from the surface and is expert at removing the whole worm from its burrow without breaking it. If you try to imitate a thrush by pulling an earthworm from its burrow with a pair of tweezers you will find that this is not as easy as it looks. Either the worms retreats rapidly, or its tip is broken off as the worm locks itself to the side of its burrow with its chaetae, hard spines which can be moved to fix the worm in position and prevent removal by an enemy.

How often earthworms surface depends on the weather. None appears when it is dry, but they are abundant during and just after rainfall. Thus a song thrush feeding its young in spring may be short of food if the weather is dry, and may have to seek alternatives such as snails, but if the weather is wet there are plenty of worms. Clearly it is not so much the abundance of earthworms but their availability near the surface that is important to the thrush. Every time a thrush takes a worm it is initiating a whole series of ecological events. Catching the worm in the first place depends on the weather, and once the worm is removed there is a slight and local reduction in the earthworm population and therefore in the rate of decomposition of humus. But predation by thrushes seems to have no appreciable effect on worm numbers and the empty space created by the missing worm is soon filled by another individual. Indeed the speed with which empty spaces are filled is one of the outstanding features of the living world, and is a problem to which much thought has been given. No one knows whether thrushes and other predators control the size of earthworm populations, but if an area becomes bad for catching worms the thrushes try elsewhere, which suggests that the availability of worms affects the abundance of thrushes.

We do not know whether the thrushes could survive if there were no earthworms but we can be certain that they would have a totally different relationship with the environment, since earthworms form such an important part of their diet. Neither earthworms nor thrushes become so common as to dominate the environment and so their populations must be balanced in some way. Earthworms, thrushes, and all other organisms have high reproductive capacities and are theoretically able to increase in numbers very rapidly. The observation that they do not do so is so elementary that its significance is often overlooked. We can conclude that a patch of grass can support a certain number of earthworms and no more, and that these in turn provide food for a certain number of predators (including thrushes), which also take other food. We can say with confidence that some sort of balance is maintained, but exactly how is another question.

## The importance of ecology

Most relationships between plants and animals and their environment are baffling in their complexity and it is virtually impossible to make

assumptions about the outcome of a deliberate change in or interference with the natural environment. This is because ecological
relationships tend to be more than the sum of the component parts.
You can take a machine to pieces and gain a fair idea of how it works
from the constituent parts and their arrangement relative to each
other, and you can with some confidence reassemble the parts and end
up with what you started with, but such an approach would be
impossible with an assemblage of plants and animals that constitutes a
natural environment. As each new component (an individual or a
species) is added to the assemblage, new properties and new attributes
are developed from the new relationships that arise. Many complex
natural environments such as tropical forest are largely self-regulating
as long as they are not tampered with; thus an alteration in one part of
the system generates compensation in another part, a balancing
process known as homeostasis.

There are probably no areas in the world, however high, deep, cold,
or barren, that are entirely free from the influence of man. It is therefore
self-consciously academic to consider ecology as something apart from
man and then assess man's impact on the 'natural' world. Rather,
man's activities from building and operating nuclear power stations to
factory farming should be considered as an integral part of the
complexity of the living world and are just as 'ecological' as a fen or a
forest. Ecology has grown from being a minor branch of biology to an
interdisciplinary study which, as the American ecologist E. P. Odum
suggests, 'links the natural and the social sciences'. Hence the special
role of the ecologist may well be to take an all-inclusive approach to the
world's problems in contrast to the approaches taken by economists
and politicians and, it may be added, many scientists.

In 1968–73 the region of Africa known as the Sahel which includes
the countries of Mauritania, Senegal, Mali, Chad, Niger, and Upper
Volta, suffered a catastrophic drought in which an estimated 100,000
people and five million cattle died. The people that suffered most were
nomadic pastoralists who move their livestock around in order to make
the most of the limited grazing available in this extremely arid region.
Under normal circumstances the pastoralists move their herds south in
the dry season and eventually reach the farms of the settled cultivators
of the south. There the animals feed on the stubble left from the
previous harvest and at the same time manure the fields with their
dung. As soon as the rainy season starts again the pastoralists drive

their herds north, continuing as long as there is green pasture ahead. This mutually helpful arrangement between pastoralists and cultivators puts no undue pressures on the land and is facilitated and administered by tribal chiefs.

A tragic coincidence of events put an end to this well-tried system of land use. First, there was the severe though not unprecedented drought; secondly, national governments in the newly independent nations of the Sahel region interfered with and in some instances stopped the pastoralists' migrations; thirdly, cultivators were induced to grow more cash crops rather than staples and to reduce the fallow period; lastly, improvements in human and veterinary medicine over a forty-year period had resulted in an increase in the human population by a third and a doubling of the cattle population to a level that the pasture could no longer support. There was outside interference, too, especially in the provision of deep water-holes funded by aid programmes, the result of which was to concentrate the cattle in certain areas to an extent that the pasture was totally destroyed by their trampling and they were left with nothing to eat.

The result of these activities was to leave seven million people dependent on hand-outs of food; they were, if you like, 'ecological refugees'. Moreover the fragile Sahel environment was badly damaged and many think that the situation now (1979) is worse than it was in 1973 when the drought ended. The tragedy could have been avoided if there had been a proper understanding of the delicate ecological balance in traditional patterns of land use: the provision of water-holes seemed so helpful but by itself it was disastrous.

The people of the Sahel would be described by many of us as primitive. Certainly they lead a life untouched by many of the trappings of the twentieth century. One of the most controversial gains during the twentieth century in Britain and other industrial countries is in nuclear technology, and ecologists, both professional and amateur, can be relied upon to contribute to the heated discussion that necessarily precedes decisions over the development of nuclear power plants. This was particularly so in 1977–8 when there was a planning inquiry into the advisability of building a nuclear reprocessing plant at Windscale in Cumbria. The opponents of the scheme marshalled evidence about the risks of radiation to workers and local residents, contamination of the land and marine environment, and the possibilities of sabotage or terrorist attack. Many appeared indignant that

*Plate 3:* Starving cattle during the Sahel drought.

Britain should be prepared to handle dangerous material that other countries feel unable to cope with, and the presiding judge was repeatedly accused of ignoring evidence that the project is dangerous and unnecessary. Is this an economic, social, political, scientific, or some other issue? We can, in any case, be sure that it has ecological connotations and the articulate ecologist can often bring together the threads of the argument. No one wants to precipitate an environmental disaster in northern England, but at the same time we must soon make provision to replace the dwindling supplies of fossil fuels upon which our society depends. At the moment nuclear power seems the best solution.

The Windscale project contrasts markedly with the Sahel disaster. It was discussed beforehand, not analyzed afterwards. People were consulted and although cynics might say that no one takes notice of what the people think there was at least the chance to protest. The people of the Sahel were never consulted despite the fact that they know more about the delicate ecological balance of the region than British citizens know about the pros and cons of nuclear power. Both incidents demonstrate the practical value of an all-inclusive or ecological approach to human problems.

## The scope of ecology

In this chapter we have introduced our subject by taking examples of ecological events induced by man and by other organisms. Many human activities produce changes in the environment, frequently of a disrupting nature, but similar events occur in the natural world where man's impact is negligible. Natural environments have a remarkable capacity to resist change, but the trouble with human activities is that they are on such a massive scale.

Leaving aside for the moment the effects of man, it appears that natural environments persist for long periods more or less unchanged. Various species of plants and animals occur year after year, interacting with each other in a highly intricate way, yet despite the enormous potential of all species for increase in numbers, most species remain relatively rare for most of the time. A professional ecologist is interested in finding out how this balance of nature is achieved and his findings are likely to be of significance to us all because they will help to suggest the consequences of upsetting the balance.

It might be argued that lawns are not especially important, but at the same time maintaining a lawn is an example of upsetting the balance of nature because a lawn would not persist unless constantly attended by man. Cutting down a vast area of tropical rain forest, the most complex environment in the world, is an ecological event of enormous magnitude, the consequences of which are poorly understood, although the result is often an impoverished environment in which few crops can be grown. There is really much in common between maintaining a lawn and cutting down tropical forest, except that the effect of one seems small while the effect of the other may be disastrous to human welfare.

A world dominated by people who are systematically destroying natural environments that have taken millions of years to develop seems to be an inescapable legacy of the agricultural and industrial revolutions. Let us hope that there will be another revolution in which attempts will be made to reduce exploitation and to conserve environments, although it must be admitted that the prospects of such a revolution seem slender at the present time. Demands for space, food, and resources are now so intense that possibilities for conservation are becoming more and more remote, but we should at least be aware of what we are doing.

From what has been said so far it will be evident that ecology is not a discrete subject and that it can be approached at several levels. In the remainder of this book we shall examine some of the more important ecological properties of living organisms and their environments, and when this has been completed we shall try and place man in an ecological framework. It is possible that some of the lessons of ecology can offer scope for planning more intelligently the kind of world that will be inherited by our children.

# 2  Thinking exponentially

This chapter explores in more detail the capacity of plants and animals to increase rapidly in numbers, often described as exponential growth.* This term, like the word ecology, has entered into everyday affairs but is frequently misunderstood. Before discussing exponential growth and other exponential processes we must be clear about what is meant by a population, an important concept in ecology to which we shall make repeated reference throughout the book.

## Populations

Many of the organisms we see about us belong to different species. It is easy to see that in the garden there are different kinds of birds and shrubs but less easy to see that the grasses in the lawn and the earthworms in the soil are not all the same. Organisms of one species breed freely with one another but not with organisms of different species. This is a good working definition that will serve our present purpose, but the species concept itself is more complicated than this, as will be explained in Chapter 4.

All people belong to the same species, as at least in principle they can breed together; in practice, however, there are geographical and social barriers which tend to inhibit large-scale interbreeding. Much the same is true of other species and so we shall use another unit, the population, to discuss what is perhaps the most important attribute of living organisms, their capacity to increase in number.

The word population is from the Latin *populus* which means people,

*Since exponential growth of populations and exponential consumption of resources figure conspicuously in various parts of the book, I feel that I should explain that I have used the word exponential broadly to mean to increase or decrease by exponents. The exponent (in my interpretation) may remain constant or may vary, and particularly when dealing with consumption it is clear that a critical threshold may be reached beyond which further consumption is unlikely or impossible. The process is not then strictly exponential and I am aware that mathematicians might quarrel with my use of terminology, but I hope I will be forgiven for speaking generally about the real world even if some of the points I make lack mathematical precision.

and for a long time virtually all discussion of population concerned human numbers. Nowadays the word population is used by ecologists to mean a number of organisms of the same species living and breeding together. It is therefore inaccurate to speak of a population containing more than one species, although unfortunately this occurs even in the more scientific ecological literature. Daisies growing in a garden during the summer constitute a population, but the same species of daisy growing in another country ten years later would to all intents and purposes form a separate population. We cannot be too rigid because in the broadest sense a population is equivalent to a species, but in a more narrow and useful sense it is convenient to think of species as being made up of a whole series of populations, each separated from the others by distance, time, and other barriers. Much depends on the kind of organism we are speaking about. Snails living in a hedgerow isolated by fields and roads from other snails of the same species constitute a distinct population, but birds (which are highly mobile) are less likely to be affected by these barriers and their populations therefore occupy larger areas.

## What is exponential growth?

If you want to save money and you distrust the bank you can set aside, say, £100 every year in what you believe to be a safe place, perhaps under the bed. Assuming that no one discovers and steals your money, your savings will grow, but the growth will be linear (just £100 a year). If on the other hand you invest it in a reputable business, the amount saved will increase at a constant percentage of the total money saved over a period of time. Interest rates on invested money vary, but figures of between five and ten per cent are common. Thus if you invest a hundred pounds a year at an interest rate of seven per cent and you leave the amount invested plus the interest in the bank or business you will have about £1400 after ten years, whereas if you had left the money under the bed you would have only £1000. Investing money is obviously better than simply putting it under the bed. Invested money undergoes exponential growth, and what is happening is precisely the same as can occur in a population of organisms.

In *The limits to growth*, a report published on behalf of the Club of Rome, exponential growth is illustrated by telling a riddle. The riddle makes some assumptions for which there is little biological justification, but this does not matter. Imagine a population of water lilies growing on the surface of a pond. The population is doubling in size each day and is allowed to increase unchecked until after 30 days it completely covers the pond to the exclusion of all other plants. For many days the lilies are quite inconspicuous, and no attempt is made to restrict their growth until they cover half the surface of the water. On what day will this be? The answer is the 29th day, the day before the lilies completely cover the pond, when there is only one day left to save the pond from being choked by lilies. This riddle shows how an exponential process seems at first to be unimportant and it is only later that its effects become conspicuous and sometimes worrying.

Yeast obtains energy for growth by the fermentation of sugar, a process yielding alcohol, which has a variety of industrial and commercial uses. The small yeast cells increase in number by dividing or 'budding' every few minutes and in a short time a few cells can grow to a population of millions. If there is an unlimited supply of sugar and if the environment is good for the survival of yeast, the cells increase in number exponentially. For instance if a population were to start with one cell its growth would proceed as follows:

1, 2, 4, 8, 16, 32, 64, 128, 256, 512, 1024, 2048, 4096, 8192 . . .

and so on until within a few hours there would be millions of yeast cells. In practice of course the sugar is always in limited supply, and the population growth eventually slows down and stops through lack of food and space, and because the environment becomes contaminated by the metabolic products resulting from growth. The growth of yeast cells looks even more dramatic if the figures are plotted on a graph as in Fig. 5. The initial slow increase in numbers is followed by a very rapid increase that causes the line of the graph to rise almost vertically.

Another way of expressing exponential growth of a population is to calculate the percentage increase in a period of time. For many plants and animals (including man) a year may be taken as a convenient period of time and an annual percentage increase calculated. This figure is not entirely satisfactory: it tends to obscure the real effects of exponential growth because the percentage may be quite small. It is therefore common to calculate another figure from the percentage annual increase, the doubling time (in years) of the population. Some examples of doubling times are shown in Table 1. If the population is increasing at only 0·1 per cent a year the population will double in seven hundred years, but if the rate of increase is seven per cent a year

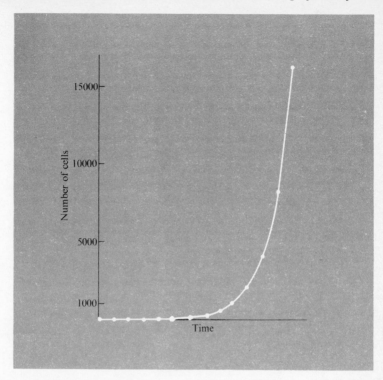

*Fig. 5:* Exponential growth in a population of yeast in which the cells are dividing every few minutes. The dots show the number of cells in the population at each time interval. At first the increase in numbers is so gradual that it cannot be shown on a graph of this scale.

*Table 1* The doubling times of populations with various rates of increase

| Rate of increase expressed as per cent a year | Doubling time in years |
|---|---|
| 0·1 | 700 |
| 0·5 | 140 |
| 1·0 | 70 |
| 2·0 | 35 |
| 3·0 | 24 |
| 7·0 | 10 |
| 10·0 | 7 |

the population will double in only ten years. The human population is currently increasing at a rate of 1·9 per cent a year which means that if the rate is maintained the world population will be twice its present size in thirty-seven years' time. A rate of increase of 1·9 per cent does not sound very great, but the prospect of there being twice as many people in thirty-seven years as there are now may be viewed with a certain amount of pessimism; to many of us there already seem to be too many people.

The rate of increase of the human population is not the same in all parts of the world. The richer industrial countries have the lowest rate of increase and the poorer nations the highest. The populations of many poor countries are likely to double in less than twenty years if present trends continue, a prospect which does not suggest much hope for their economic development. Long-term educational, medical, and social planning relies on accurate forecasts of the birth rate. In Britain the number of births was until recently expected to continue declining up to the 1980s, but in the first quarter of 1978 births increased for the first time for several years. If this rise is maintained it may have far-reaching consequences: just how many doctors, nurses, and teachers are we going to require in ten or twenty years' time, and should we increase or decrease the number of people being trained in these and similar professions? Even in a country like Britain there are variations in the rate of population increase correlated with social class, wealth, religion, and race. If, for example, the rate of increase in the immigrant population of a city is 3·0 per cent a year while the rate for residents is only 0·5 per cent, there will be twice as many immigrants in 24 years, but we have to wait 140 years before the resident population is doubled. Authorities tend to avoid giving figures of population growth in relation to racial origin on the grounds that we should look towards an integrated society, but suppression of information, always danger-ous, can easily lead to suspicion and exaggeration of the true state of affairs. Thus in 1978 the Secretary of State for Social Services in the British Government accused another member of Parliament of doing great damage to race relations by his 'obsession' with the birth rate of immigrants. The member of parliament in question had suggested that a third of births in Britain were to mothers from the New Com-monwealth and Pakistan. As a consequence figures were issued for the years 1969–76 showing births to mothers of immigrant origin in five cities and the boroughs comprising the inner London area; births to (immigrant) mothers comprised twenty-two per cent of the total. The 1977 Demographic Review shows that although the average number of children born in Britain to Indian and West Indian women in 1971 was significantly higher than the average for the indigenous population, by 1976 there had been a reduction in family size among West Indians but not among Asians. Population growth is a vital issue, and all citizens should be entitled to the maximum possible information. It is only in this way that suspicions can be averted and people can judge for themselves whether this or that policy is workable or desirable.

An all too familiar example of exponential growth is the cost of living. Everyone knows that rising costs constantly devalue money and despite much political intervention inflation continues as a seemingly unending process. In many industrial countries there is at the moment an average increase in the cost of living of about five per cent a year which means that food and other essentials will cost twice as much in fourteen years' time as they do now. Rising costs lead to demands for higher wages and there is no obvious prospect of stability. Inflation is one of the social consequences of rising standards of living, and as with population growth it cannot be stopped unless there is a drastic reduction in the availability of resources, which in turn would stabilize and even reduce the standard of living of people in industrial countries.

## Reproduction and death rates

The rates of reproduction of many organisms are extremely high. As mentioned in Chapter 1 some fish lay hundreds of thousands of eggs and other organisms, like oysters, produce millions of offspring every time they breed. Since the world is not full of fish and oysters, or of any other organisms, it follows that death rates are very high indeed; often more than ninety-nine per cent of the eggs, young, or seeds produced fail to survive, and many of those that do survive die before they are able to reproduce themselves. When the recruitment of individuals into a population and the death rate are equal there is stability – zero population growth – and it is only when recruitment exceeds deaths that the population increases.

In man the birth rate and death rate are usually expressed as the number of babies born and of people dying in a year per thousand of the population. In the world as a whole the current birth rate is about 34 per thousand and the death rate 15 per thousand. Every year, therefore, the population increases by 19 per thousand, or 1·9 per cent. In mid-1969 the world population was estimated at 3551 million, and so we can estimate that there will be 4386 million people by 1980. But the rate is liable to alteration, up or down, and it becomes necessary constantly to revise our projections, and there may be more or fewer than the projected number by 1980. It must also be noted that growth of the human population results in exponential consumption of resources, which include not only food, but also non-renewable raw materials like oil and coal.

Few populations occurring in nature undergo long-term exponential growth in size, and indeed almost all the known examples are in situations where man can be implicated as having an influence on what is happening. In most organisms births and deaths are approximately balanced and although most populations fluctuate within limits they remain on the whole relatively stable.

## Examples of exponential population growth

For centuries man has been introducing plants and animals into parts of the world where previously they were absent. These introductions often fail, but some are successful, and the consequences may be disastrous for native species. Islands seem particularly suitable for the successful introduction and establishment of plants and animals from elsewhere, possibly because islands tend to contain fewer species than continents and are therefore more receptive. In contrast undisturbed tropical forest is resistant to introductions, possibly because of the large numbers of species already present. The spread of an organism in a new area is slow at first, but soon gathers momentum in much the same way as the increase in yeast cells shown in Fig. 5.

There were several unsuccessful attempts to introduce the European starling into North America and it was not until 1891 after eighty birds had been released in Central Park, New York, that several pairs bred. From then on the bird spread, rather slowly at first and chiefly in and around cities, the population suffering a severe set-back in the cold winter of 1917–18. But by 1954 it was found almost everywhere in the United States and had penetrated into Alaska and Mexico; there are now millions of them. The starling has also been established in South Africa, Australia, and New Zealand, and on many small islands; even in Europe its numbers and range have increased enormously as the result of expanding agriculture and the development of towns and cities, which provide environments favourable to the species. Starlings provide a classic example of a population explosion brought about by man, and there is no doubt that the increase both in numbers and in geographical range has been generated by human alteration of the environment as well as by deliberate introduction.

Many other European birds have been introduced into different parts of the world, the most successful species being the house sparrow, the partridge, and various kinds of pheasant, which however are themselves introductions to Europe dating back several centuries. Other species, notably the fulmar which has become a scavenger around fishing boats and harbours, have increased dramatically because of human activities without having been deliberately introduced. Fulmars are gull-like sea-birds and spend most of their time over the open sea feeding on small animals living near the surface of the water. They are now probably the commonest sea-birds in the North Atlantic. Fulmars breed in the spring on cliffs, and until 1878 the sole breeding place in Britain was the tiny island of St Kilda. Since then they have colonized almost the entire British coast wherever there are suitable nesting sites, even as far south as Devon. Everything we know about the fulmar's success as a species suggests that by changing its feeding habits and becoming associated with fishing and whaling it has been able to build up in numbers to become one of the commonest birds in the world.

Some of the commonest weeds are introduced species. North America and New Zealand are full of European weeds and in Africa a daisy-like flower called *Tridax procumbens*,* first introduced from Central America, has now spread right across the continent, colonizing bare ground created by agriculture and road-building. Oxford ragwort, at one time a rare plant in Britain, first spread from the City of Oxford (where it was being cultivated in the Botanic Garden) along railway lines, colonizing waste land near railway stations, and later penetrating into towns. It is now one of the most abundant weeds in Britain. Equally dramatic has been the spread of the rose-bay willowherb, whose purple flowers are now so much a part of the British countryside that it is hard to believe that it was until recently quite a rare plant. This species was one of several plants able to exploit the new environments created by bombing during the Second World War.

* Latin names for species are used in this book in cases of ambiguity or where there is no suitable English name.

Since then it has colonized almost every patch of waste land in the country, being especially abundant on building sites and places where trees have been felled. In Scandinavia it has spread along roads well north of the Arctic circle, but it does not seem able to establish itself in undisturbed areas. Many weeds and the insects that feed on them have been accidentally introduced, and although it is common to assume that cultivation and building lead to the disappearance of species of plants and animals there are a great many species which are able to exploit the changed circumstances.

A large snail, *Achatina fulica*, growing up to 20 cm in length, lives on the East Africa coast where it is not especially common. It was introduced to Madagascar at an unknown date and from there it spread to many islands in the Indian Ocean, reaching Mauritius by 1800, Reunion by 1821, and the Seychelles by 1840. Before the end of the nineteenth century it was well established on the mainland of tropical Asia. It then colonized islands in the Far East, reaching Japan by 1925 and later spreading to the Pacific islands, reaching Hawaii in 1936 and California by 1947, where, however, possibly because of the cool climate and control measures, it did not become well established. *Achatina fulica* seems to have been spread accidentally in packages shipped from one place to another and its ability to survive for long periods without food or water no doubt contributed to its success as a colonizer, but it was also spread deliberately as a source of food. Some Japanese troops are reputed to have carried young snails around and introduced them to islands where they were stationed during the war. Thus in about 150 years this species extended its range far beyond its original home in East Africa, but curiously enough it does not seem to have spread within Africa itself, possibly because of the presence of other similar species there. In many places it has become an agricultural pest and various methods of control have been attempted, with only limited success. One problem is that poor people, although concerned about the snail's destruction of their crops, use it as a source of food – a good example of how conflicting interests can be detrimental to schemes aimed at controlling pests.

As mentioned earlier islands are particularly liable to colonization by species introduced by man, and two groups of islands have received an extraordinary variety of alien plants and animals. In these islands, Hawaii and New Zealand, the native species have become much reduced in numbers and some species are now extinct. It is not known for certain in every case whether introduced species have directly caused the extinction of native species, for it is likely that man's alteration of the environment has contributed, but in some cases there seems little doubt that introduced species have displaced native ones by taking over their resources. Before the arrival of Europeans there were no native land mammals (except bats) in New Zealand, but by 1950 about thirty species were well established, many of them regarded as detrimental to the local flora. One species, the red deer, increased so rapidly that it has damaged the natural vegetation and caused severe soil erosion. Many of the common birds and mammals of modern New Zealand were introduced from England. They include such familiar species as hedgehog, stoat, brown rat, black rat, rabbit, hare, mallard, skylark, song thrush, blackbird, rook, chaffinch, redpoll, goldfinch, and greenfinch, to name just a few. Other birds and mammals were brought from North America, Asia, Australia, Polynesia, and western Europe, and there have also been successful introductions of insects and plants. Many were deliberately introduced by Europeans who wanted to have around them familiar plants and animals as a reminder of home; many introductions failed, but of those that were successful the pattern of spread was the same: a slow start and then a rapid increase, sometimes so rapid that the plant or animal became for a time far more abundant than in its original home.

Even more spectacular population growth occurs in disease organisms during an epidemic. People in all parts of the world are constantly alert to the threat of epidemic diseases like cholera, bubonic plague, smallpox, and influenza, all well known for the speed at which they can spread through the human population. Influenza is caused by a virus and the disease repeatedly sweeps through vast areas of the world affecting millions of people, eventually to disappear as suddenly as it appeared, but returning again perhaps in a year or so in a new and initially more damaging form. Each of the various kinds of influenza is a mutant virus to which at first people are not immune, and almost as soon as one mutant form disappears another is on the increase somewhere else. The rapid spread of disease organisms means that once the disease has a firm hold there is little that can be done to stop it. It is estimated that in the First World War about a hundred million people died of influenza, which provides some idea of the explosive nature of the disease once it has established itself.

All the examples of exponential population growth considered thus far are based on evidence of geographical expansion and not on the actual numbers of organisms involved. This is because it is by no means easy to assess the numbers of most organisms over a period of time, especially when you consider that the early stages of an exponential process are not obvious, and it is only when the organism has a firm hold that people become aware of what must have been happening. This partly explains why people rarely think about the consequences of deliberately introducing a plant or animal into a new area and why many diseases go undetected until there is a large outbreak. But occasionally it has been possible to follow the growth of a population after the initial introduction, as in the next example.

Pheasants originated in Asia, but are so familiar in many parts of Europe and North America that they are likely to be taken as indigenous. Several species have been spread around the world by man so as to provide sport and food, the most abundant being the common or ring-necked pheasant, *Phasianus colchicus*, a bird much beloved by the shooting fraternity in all countries where it is established. Pheasants are afforded protection and encouragement, especially in the breeding season, and much effort and money is devoted to their preservation. In the north temperate region they breed in the spring and lay a clutch of 8–15 eggs. The young birds are fed and tended by the female and fly in about 13 days.

In 1937 two males and six females were introduced to Protection Island off the coast of Washington in the north-west of the United States. There were no pheasants on the island until the introduction, but the environment must have been suitable for them as the population rose from 8 to 1898 in a matter of six years. The population was counted each spring and autumn and the numbers recorded are plotted in Fig. 6. Initially the population increased slowly but it soon gathered momentum in the now familiar pattern. As shown in Fig. 6 there were always rather fewer pheasants in the spring count, no doubt because of mortality during the previous winter, and because of this the population rose in a series of steps. No figures are available after 1942, but already the rate of increase was showing signs of slowing down. Similar counts of pheasants in areas where they are already well established show the same fall in numbers each spring, but there is no long-term growth in the size of the population: numbers fluctuate in relation to the severity of the winter, shooting pressures, and the

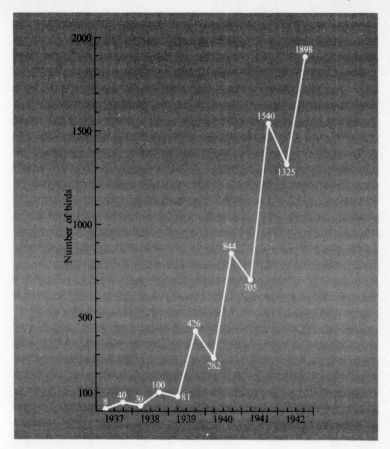

*Fig. 6:* Increase in numbers of the pheasant on Protection Island, Washington. A census was made each spring and autumn. (From D. Lack, 1954)

availability of food to the young birds in summer. The growth of the pheasant population on Protection Island is similar to that of the yeast cells shown in Fig. 5.

## Population decrease

Fossil evidence shows that many species of plants and animals once common on earth are now extinct. No one really knows why organisms become extinct but it seems that a species may flourish for a time, perhaps for millions of years, and then there is a sudden reduction in its numbers, followed by a period of rarity until it finally disappears. Some species are known to have disappeared more suddenly than others but it would seem that decline and eventual extinction follow in reverse the pattern of population increase discussed in the previous section.

Most people have visited museums and have been impressed by the fossil remains and reconstructions of extinct dinosaurs. These animals first appeared about 190 million years ago and for millions of years they were the largest and most conspicuous of the land animals. Some weighed up to 50 tons, others were small, no larger than a chicken. There were many species, the largest being plant feeders, while many of the smaller species were carnivores. Dinosaurs have been extinct for about 80 million years and their nearest surviving relatives are the crocodiles and the birds; indeed it could be argued that birds have replaced dinosaurs. Individual species of dinosaurs became extinct quite quickly but as a group their disappearance was a long process, species after species dying out slowly over periods of millions of years. We can only speculate as to what caused their extinction. By using a certain amount of imagination we could draw a graph of dinosaur history indicating a slow build-up followed by a period of great abundance and then a decline showing the features of exponential growth in reverse.

Something similar to what must have happened to the dinosaurs is replicated whenever man attempts to control a pest or weed by the application of pesticides or herbicides. If for instance a garden is over-run by one of the persistent weeds such as the sorrel, *Oxalis*, an attempt may be made to eradicate it. The first application of the appropriate herbicide will appear to kill virtually all the plants and the gardener may feel that he has succeeded at the first try. But the chances are that some plants will reappear after a short time and if no further measures are taken the weed will spread over the garden again and soon reach its former abundance. One way to combat this possibility is to apply the herbicide repeatedly until the weed is finally extinct in the garden, but this may be a long job, the weed showing remarkable and annoying abilities to survive and re-establish itself. The analogy between eradicating *Oxalis* in a garden over a period of perhaps one or two years and the extinction of the dinosaurs over millions of years should not be taken too far, for apart from other considerations the destruction of *Oxalis* in a garden does not result in its extinction as a species, and there is always the chance that the garden will eventually be colonized again from elsewhere. But there are similarities, the most important being that there is initially a big reduction in numbers followed by a long period during which the organism is rare. The length of this period will vary with the circumstances and a really efficient herbicide applied repeatedly will quickly destroy the entire population of a weed without difficulty.

Consider as a further example a heated greenhouse in which attractive plants are being grown. The plants may become infected with one or more species of red spider mite, a pest feared by greenhouse owners. The mites increase exponentially and continue to do so until severe damage is done to the plants. The greenhouse may then be fumigated with a chemical toxic to spider mites. Most if not all of the mites will seem to disappear quickly, but some will survive and the population could easily build up again to its former level. It therefore becomes necessary to apply control measures repeatedly if the mites are to be exterminated, and this may be a long and tiresome process.

Gardeners and farmers will be familiar with the events described above. Medical entomologists trying to eradicate insects like mosquitoes which transmit disease are faced with a similar problem. It is possible almost to eradicate a mosquito population by draining the swamp where the insects breed and by spraying the area with insecticides, but unless there is a follow-up period during which control measures are repeated the work is liable to lead to frustration and despair as numbers build up again.

And so while exponential growth of a population can lead to unexpectedly large numbers in a short time, exponential decrease in numbers (which is exponential growth in reverse) can lead to a superficial impression that a population has disappeared. If we are dealing with a weed or a pest caution must be exercised as the chances are high that the population will recover. It must be admitted however that the causes of a population almost disappearing and then recovering again are in most cases obscure, just as obscure in fact as the causes of extinction.

### Thrushes eating snails

If a source of food exists in a fixed quantity, neither on its own decreasing nor increasing, and its consumers are constant in number, their consumption of the food will follow the reverse of an exponential growth curve. At first the rate of consumption is high but as the food is used up the rate falls and when it has nearly gone the rate may be very low indeed.

It is relatively easy to show by experiment that rates of consumption vary with the quantity of the resource available. In the following experiment the resource is a population of snails and the consumers are song thrushes. The period of the experiment was short enough that there was increase neither in the number of predators nor in the number of prey.

The song thrush is adept at feeding on snails, especially the larger species found in gardens and hedgerows. Among those commonly eaten are the garden snail, *Helix aspersa*, and two species more often associated with rural areas, *Cepaea nemoralis* and *Cepaea hortensis*. The song thrush takes snails mainly when the weather is dry and when earthworms (its favourite food) are hard to find. When a thrush finds a snail it takes it to a stone and smashes the shell so as to expose the soft edible body of the snail. The same stone is used time and time again and bird-watchers sometimes refer to the stones as 'thrush anvils'. By collecting the empty snail shells from around the anvils it is possible to find out how many snails are being eaten (a discovery which is not particularly useful unless you know how many snails are available to the thrushes).

In May 1972 a sample of 724 *Cepaea nemoralis* was collected from a locality in Sweden where there are no thrushes and brought to a garden in the English Midlands where there are several resident thrushes. The snails were released on 23 May and immediately began to disperse into the vegetation, but being snails they did not move very far. There were previously only a few *Helix aspersa* in the garden, and the nearest *Cepaea* localities are so far away that it is unlikely that the thrushes had encountered them before. The thrushes first found the snails on 27 May and on that day consumed 90, or 12·4 per cent of the total. From then on the birds found snails every day and within sixteen days they had accounted for 630, or nearly 87 per cent of the total. Table 2 shows the number of snails eaten each day, the number remaining from the original total, the percentage of the original, and the percentage of the available snails eaten each day. There are several ways of looking at Table 2. First, the thrushes took more snails at the beginning of the experiment when they were most abundant. In the

*Table 2* Snails eaten by thrushes on successive days

| Date (1972) | Number of snails present at start of each day | Number eaten | Per cent of original total eaten | Per cent of available snails eaten |
|---|---|---|---|---|
| 27 May | 724 | 90 | 12·4 | 12·4 |
| 28 | 634 | 113 | 15·6 | 17·8 |
| 29 | 521 | 98 | 13·5 | 18·8 |
| 30 | 423 | 71 | 9·8 | 16·8 |
| 31 | 352 | 119 | 16·4 | 33·8 |
| 1 June | 233 | 52 | 7·2 | 22·3 |
| 2 | 181 | 51 | 7·0 | 28·2 |
| 3 | 130 | 16 | 2·2 | 12·3 |
| 4 | 114 | 10 | 1·4 | 8·8 |
| 5 | 104 | 5 | 0·7 | 4·8 |
| 6 | 99 | 0 | 0·0 | 0·0 |
| 7 | 99 | 1 | 0·1 | 1·0 |
| 8 | 98 | 1 | 0·1 | 1·0 |
| 9 | 97 | 0 | 0·0 | 0·0 |
| 10 | 97 | 2 | 0·3 | 2·1 |
| 11 | 95 | 1 | 0·1 | 1·1 |
| 12 | 94 | 0 | 0·0 | 0·0 |
| 13 | 94 | 1 | 0·1 | 1·1 |
| 14 | 93 | 0 | 0·0 | 0·0 |

first five days they took nearly seventy per cent of the original total and thereafter the number of snails taken fell until by the end of the experiment they were finding very few snails indeed. The last column of Table 2 shows that during the first week the thrushes took each day between twelve and thirty-four per cent of the available snails, and that thereafter the proportion of available snails taken fell abruptly until by the end of the experiment the thrushes were finding few snails even though there were still ninety left. This result shows that thrushes are most efficient when the snails are abundant and that as the snails become scarcer the thrushes find proportionately fewer of those remaining, but we must also bear in mind that as time passed the snails may have become better at concealing themselves, or that the snails which hid most effectively were the ones that survived.

A further thirteen snails were eaten between 13 June and 15 November, and of course it is also likely that a few died from other causes during this period, but a year later in June 1973 there were still a few live snails in the garden, although the thrushes had long since stopped actively searching for them. In the end the introduction was successful, the snails now reproduce every year, and by 1978 there were probably about twenty descendants, well scattered through the garden; just occasionally one is found and eaten by a thrush. Evidently the garden is capable of supporting only a very small population of *Cepaea nemoralis* and there are indications that thrushes are responsible for keeping numbers low.

The experiment shows that as a resource becomes scarce there is less possibility of it being exploited. In this case the thrushes had available alternative food like earthworms and insects but had there been no alternatives the situation for the thrushes would have deteriorated and they would have had to move elsewhere or suffer from food deprivation.

## Exponential consumption of resources

If a population is increasing exponentially there will be a tendency for the resources upon which it depends to become depleted exponentially. This tendency will be resisted if the resource is renewable: under normal conditions grass eaten by cows is replaced by the growth of more grass. But if the resource is in fixed quantity and not renewable it will be depleted as the demands of the population increase. In the example just given of thrushes eating snails the resource was potentially renewable but in this instance the thrushes depleted the snails before they had a chance to breed and reproduce themselves. This does not normally occur and in general the demand for a resource is balanced more or less precisely by the supply. Effectively all organisms except man are dependent on resources which are renewable and in this section we shall touch briefly on the special position of man as a consumer of non-renewable resources. The question of the consumption of renewable resources and the resultant stability of populations is taken up in the next chapter.

Like other animals man's chief resource is food which is potentially renewable through the growth of plants. Virtually all of man's food is derived directly or indirectly from plants but some of the energy used to obtain this food is obtained from sources that are not renewable. The most obvious of these are the fossil fuels (coal, oil, and natural gas) but we should also include minerals like iron, copper, zinc, tin, and aluminium, which are used in the manufacture and operation of machinery for agricultural work. It would theoretically be possible to recycle metals and there is an increasing demand that this should be done, but large quantities become more or less locked up in buildings and machinery and are effectively lost so far as future use is concerned. There is also an enormous wastage, which although unnecessary is difficult to control: it is still cheaper to use new raw materials than to recover materials already used. But the fossil fuels are in a different position because their stored energy is lost in combustion and cannot of course be recovered. Man is therefore dependent on an important source of energy which occurs on earth as a fixed quantity and once it has gone there will be no question of renewing it.

Man's consumption of non-renewable resources is increasing exponentially. For example, from what is known of remaining world reserves, lead will to all intents and purposes be unavailable as a raw material some time in the 1980s. Other metals which may become unavailable as raw materials before the turn of the century include silver, gold, mercury, platinum, tin, and zinc, while copper, which has many important industrial uses, is expected to run out by about the year 2100. The exploitation of these and other metals involves the conversion of raw materials into structures and machinery required by our industrial society. In a sense these metals are never lost and most are potentially recoverable, but recovery will not occur until the price of raw materials rises so high as to make it profitable. Man has re-distributed naturally occurring materials and has locked up vast quantities in his industrial enterprises – he has converted part of the natural world into a man-oriented world. Thus when we say that the supply of a metal is running out this simply means that we are approaching a point where further exploitation is unprofitable. In some instances improved technology makes it feasible to consider exploiting low-grade deposits of minerals, but at enormous cost both financially and in terms of the scars left on the countryside. The beautiful but diminishing tropical rain forest of the Freetown penin-sula in Sierra Leone is threatened by the presence of platinum in the rocks beneath it; indeed tropical forests the world over may be so devastated by mining operations that the trees will never establish themselves again. Many factors other than supply and demand affect the availability of non-renewable resources to the world at large. Zambia, not a rich country, depends on the sale of copper on the international market for almost all its export earnings. A fall in the world price of copper leaves Zambia spending more to extract the metal than can be obtained when it is sold. Needless to say Zambia's economy is tottering, although the British Government helps, with total expenditure in capital aid and technical co-operation for 1977–9 expected to be about £47 million. Zaïre depends on the mineral-rich Shaba province (formerly Katanga) for 75 per cent of its export earnings; Shaba supplies 6 per cent of the world's copper and 50 per cent of the cobalt required by the expanding aerospace industry. The recent (1978) insurrection in Shaba has frightened away many of the European technicians who run the mines; what this will do to world supplies of cobalt cannot at present be predicted. A cubic kilometre of sea water (which weighs $10^9$ tons) contains many metals, including 1·3 million tons of magnesium, 300 kilograms of silver, and 4 kilograms of gold, but the extraction of silver and gold and other metals from sea water is not a viable proposition. If the rate of consumption itself increases, many metals may become effectively unavailable sooner than current predictions would suggest. During the final stages of exponential demands we can expect prices to soar. The last to suffer will be those who are already the biggest consumers, as these people are in the richest nations who alone will be able to afford the cost of raw materials. Poor countries which so far have been able to use resources in only small quantities will never have a chance to establish competitive industries.

Virtually everything written about the rate of exponential consumption of raw materials is out of date by the time it is published. At the time of writing the first edition of this book, there was a report in *The Times* of London dated 13 March 1973 which was headed 'Indicators reveal unprecedented surge in costs of raw materials'. The report goes on to say that in one month – February 1973 – the cost of basic materials and fuel used in the manufacturing industries in Britain jumped by another three per cent. Raw materials increased in price by about twenty-one per cent in a year, which, as *The Times* remarks, is an astonishing rate of inflation, much greater than in previous years. And yet this is precisely what is to be expected: as human numbers and human demands grow there will follow an exponential depletion of resources and rapid rise in costs. Indeed the universality of exponential processes has its sinister aspect: big multi-national companies are increasingly taking over world trade and small companies are disappearing, which means that big business will establish a tighter hold on governments. If we are not careful our destiny will be determined by the pressures and wishes of the big companies who alone will be able to pay the prices for the ever-diminishing capital stocks of raw materials essential to industry.

Economists sometimes define economic development as a cumulative process which increases net consumption, and a common measure of development is the average income per head of the population, but in ecological terms income per head is a measure of resource consumption. Economic development can therefore be seen as an exponential process to which most politicians and businessmen are firmly committed but which like all exponential processes cannot be expected to go on indefinitely.

Economic development also results in increased pollution of the environment. We can therefore expect pollution to increase exponentially. The real dangers from pollution are not yet with us but we can expect them to occur with the same dramatic suddenness as when on the 30th day the lily population covered the remaining half of the surface of the pond.

Oil is now the most important of the non-renewable resources utilized by man and has replaced coal, which in the nineteenth century provided the means for the industrial revolution. Modern industry is absolutely dependent on the ready availability of oil. At present well over half the world's supply of energy for industry and mechanized farming comes from oil and without it vast numbers of people would suffer deprivation and many would simply starve. As recently as 1926 at least eighty per cent of the world's energy came from coal, but oil has now been substituted because it is easier and more profitable to produce and because it is (or was) believed to be cleaner and less polluting.

In the first half of 1978, major oil tanker disasters have raised justifiable fears about the means by which oil is transported. The grounding and subsequent break-up of the Amoco Cadiz on the shores of Brittany polluted the sea, smothered beaches in black slime, killed thousands of sea-birds, endangered the tourist trade, and ruined local shellfish harvests. By June 1978 the clean-up operation had cost £41 million. The disaster raised questions about the density of shipping in European waters, the proximity of shipping lanes to the coast, the relevance of salvage laws framed early in the present century, the assumption that ship-owning and operating is simply a commercial activity with no social responsibility, the large quantities contained in one vessel, and the inevitability of human error. These questions were

*Plate 4:* Oiled guillemot.

repeated when later in 1978 the Eleni V broke up and released most of her oil off the Norfolk coast. 1977 was the second successive year in which there was an increase in the number of incidents of oil pollution around Britain; but the source of pollution was identified in only a third of the 642 incidents, and so local authorities usually had to bear the cost of clearing up the mess. Nearly half the incidents involved pollution at sea, more than a quarter were spills at ports, and the rest caused shore pollution which in one case out of nine extended over more than a mile (1·6 kilometres) of coastline. There is every reason to suppose that we shall experience more and even worse pollution from oil in the next few years.

Once the existing oil reserves are gone they will not be replaced. It is by no means certain when the world's supply of oil will run out but there seems considerable agreement that it cannot be expected to last for more than another sixty years. All estimates depend on assumptions made about the likelihood of discovery of new oilfields, on future rates of consumption, and on the possibilities of switching to other sources of energy in the next twenty years. There is every chance that oil will be effectively unavailable for most purposes by the end of the present century. On the basis of the rate of consumption in 1973, it was suggested that although only 12·5 per cent of the world's oil reserves would have been used by 1975, by 1990 demand would exceed supply.

It now seems that some of the projections about oil resources given in the early 1970s were over-pessimistic. New oilfields are regularly discovered in the North Sea and in many other parts of the world including Africa. This has lulled us into complacency, but the underlying problem remains: oil resources are finite and consumption is increasing exponentially. Despite enthusiasm and publicity, progress in the development of alternative energy sources is depressingly slow. As mentioned in Chapter 1 there is widespread resistance to large-scale use of nuclear power because of the risk of accident or sabotage. Research continues on wind-, wave-, tide-, and water-power, but suggestions that Britain's profits from North Sea oil should be invested in alternative energy supplies meet with less than enthusiastic reception. One socially acceptable field where advances are evident is in the production of hydro-electric power. The Central Electricity Generating Board are constructing a highly imaginative (and well-hidden) complex in North Wales which will pump water to a lake from which it can be released at peak periods to drive turbines that will supply the National Grid. A similar installation is operating in Scotland. But it seems that the real motivation and finance to develop electrical cars, safe nuclear power stations, and devices that harness natural forces will only arise at the last minute when deprivation is imminent and time too short to do the job properly.

President Carter of the United States launched new energy policies soon after gaining office, but his attempts to implement them have met with much opposition, despite a vocal minority who organized the celebration of 3 May 1978 as Sun Day, a day devoted to generating public support for harnessing the sun as an energy source. In 1977 solar heating equipment to the value of £66 million was sold in the United States and the figures could double in a few years, especially if more towns follow the example of Davis, California. Davis calls itself 'an energy conserving city' and has enacted strict building regulations about window-size and orientation and heat recycling. Solar heating is expensive to install and like all alternative energy programmes costly research is needed. Who will pay? The Sun Day people are opposed to nuclear power because of its potential dangers, and they also fear that increased use of coal will lead to accumulation of atmospheric carbon dioxide, which traps warm air and results in a rise in temperature – the so-called 'greenhouse effect'. But it is on coal that one of the biggest American oil companies is relying in its well-funded attempts to supply an alternative energy source. Coal is expected to supply 27 per cent of America's fuel needs by 1990 as opposed to 18 per cent now, and the company is investing in processes that convert coal to gas and liquid fuel. The same company also anticipates that by 1990 nuclear power will supply only 10 per cent of American needs and solar, hydro-electric, and geothermal power only 3 per cent. But coal supplies are also finite, and if consumption increases exponentially supplies might not last for more than another hundred years. In the end it looks as if we shall have to use natural or nuclear forces to provide energy.

Whether or not he can power his machines, man must eat. With good management food resources are potentially renewable, although land can easily become permanently degraded by over-grazing. Marine fish are culled, not farmed, and twenty years ago marine biologists were speaking in terms of inexhaustible harvests from the sea, urging people and governments to make more use of fish as food. But things are different now: world fisheries are suddenly in a state of crisis and stocks have been so drastically depleted that fears are expressed as to whether they will ever recover. The crisis has threatened for more than ten years but has become acute in only the last three. Although predictable and avoidable it was precipitated by the very interests that had the information and influence to prevent it: no one should ever claim that a resource is inexhaustible because to do so suggests a failure to understand the nature of exponential processes. The previously rich fisheries of the European continental shelf are now in a sorry state. If present rates of fishing are maintained the number of cod and haddock taken is such that those that remain will be too few to replace the stock, which already shows signs of over-exploitation. The herring population has also been over-fished and the catch has dropped to a fifth of what it was; the smaller mesh nets now used by the trawlers take smaller and younger fish, equivalent to using capital as well as interest, so that there is little or no recruitment to the population. Since 1950 the total yield from the European continental shelf has increased but the species composition of the catch has changed. Increasingly fish that hitherto had little commercial value are caught to be processed into fish-meal for animal feed. International acceptance of 200-mile coastal fishing zones has highlighted the problems of nations maintaining distant-water fishing fleets. Currently Britain has placed a complete ban on herring fishing off the west coast of Scotland because the Common Market cannot agree on conservation measures.

As home stocks are fully exploited and economic zones zealously guarded by coastal states, fishermen look further afield. Fishing nations converge on the rich fisheries that still exist in regions of upwelling of nutrient-laden waters, as for example off the west coast of Africa. Here boats from Europe, Korea, Japan, and the Soviet Union compete effectively with local fishermen using inferior vessels and equipment. In 1977 the rich countries were taking 70 per cent of the catch from tropical and sub-tropical waters. The African states bordering some of these waters are deficient in protein food but are unable to finance modern fishing boats and factory ships. Ironically, much of the catch by boats from rich countries is destined for fish-meal and consequently is taken non-selectively, rapidly, and carelessly with deleterious effects on the stock. Already the sardine fishery off Morocco has suffered to an extent that the catch per unit effort is becoming uneconomic.

Oceanic upwellings in the eastern Pacific, especially off Peru, support vast populations of anchovy. From small beginnings in 1955 the catch grew to over 13 million tons in 1970, but then declined dramatically and in 1973 was less than 2 million tons. The primary cause seems to have been over-fishing, and although through the imposition of stricter limits it has been possible to effect some recovery the situation is at best precarious. Small wonder that countries bordering rich tropical seas are following the example of wealthier countries and establishing 200-mile fishing limits; but whether they can successfully enforce these limits is another matter.

What has happened to the great whales is even more depressing than what is happening to marine fisheries. The right, the hump-back, the bowhead, the grey, and blue whale were hunted to near-extinction in the last century by a once-lucrative industry supplying whalebone for corsets; oil for cosmetics, the leather industry, lub-ricants, and margarine; and food for people and (especially) for pet dogs. In 1874 eleven whaling ships were based at Dundee and landed 190 whales; by 1912 there was only one whaler which caught no whales. Antarctic catches of blue and fin whales totalled nearly forty thousand in 1930 but subsequently declined, gradually at first, then more rapidly until by 1967 only 2155 fin whales were caught and no blue whales. In 1954 an overall limit to the Antarctic catch was agreed and now the International Whaling Commission sets a limit on the global catch, weighted by the rarity of each species. Implementation of the limit depends on co-operation and on accurate reporting of the catch, but this stricture is not always observed, especially as eight whaling countries are outside the 17-nation Commission. Britain stopped whaling in 1963 and the United States in 1971; a ten-year moratorium was recently requested by Panama but then withdrawn for no apparent reason. Japan and the Soviet Union still kill thousands of whales each year and resist attempts by other nations to reduce the global quota. Japan has had to reduce the labour force employed in the whaling industry from a million to 200,000 people but claims that it is 'forced' to hunt for food even though the Japanese receive less than one per cent of their protein food from whale meat. More significantly Japan managed to export 3,365 tons of sperm whale oil in 1977; in other words it is not the need to feed people but profit that really counts. All eight species of great whale are on the danger list and other, smaller whales may soon be in the same position if Japan implements the possibilities it is exploring for replacing whale meat with dolphin meat. Large marine mammals with low birth rates, living in remote seas, and ill-understood in terms of ecology and behaviour are vulnerable to exploitation and hence become a non-renewable resource subject to all the dangers arising from exponential consumption.

## Available land is limited

Our use and abuse of land is similar to our exploitation of the seas. The area of land is finite, placing an absolute limit on numbers of people in terms of space they occupy and space they need to produce food. Much of the land is not capable of sustained cultivation because it is too cold, too steep, too dry, or even too wet. We often assume that a fixed proportion of the land surface is suitable for cultivation, or that through technological innovation more land will be made available. The truth is that we over-exploit the land just as we over-exploit the sea. Some regions that were once green and fertile are now unproductive desert. Mareotis, an area of rich vineyards in ancient Egypt, suffered neglect and mismanagement under Roman rule and is now dry and barren. The ancient kingdom of Ghana, once wealthy and fertile, is now lost beneath the Sahara. But the mistakes that caused degradation of land are not just past history; they are repeated today, and desertification is a serious threat in many parts of the world.

There are about 9 million square kilometres of desert and another 48 million square kilometres are at high risk, more than a quarter beyond recall. About a third of the land surface of the world is desert or in danger of becoming desert. Much of this is marginal land with fragile soil and sparse vegetation receiving little rain and suffering long periods of drought.

Expanding human populations make increasing demands on the land: trees are felled for firewood, land is cleared for cultivation and then cropped too heavily thus exhausting soil fertility, and cattle trample and damage the soil and overgraze the vegetation. Mechanized farming for cash crops demands the clearing of large tracts of natural vegetation; the consequences are increased erosion by wind and rain, and local climatic changes that may lead to a reduction in rainfall. In Sudan's Kordofan province mechanized farming of groundnuts, welcomed at first, has produced such deterioration of the land that where farmers produced 0·95 tons per hectare in 1961 only 0·21 tons per hectare could be produced in 1973.

Irrigation has been used from ancient times to allow cultivation throughout the year in dry regions. It works well if underground drainage is good as in the Nile delta, but where drainage is poor it gradually raises the water table and the soil becomes water-logged. The Sumerians irrigated the delta between the Tigris and the Euphrates with copious quantities of water of rather high salt content. About 3000 years ago the salt content of the soil increased to a point at which they were forced to change from growing wheat to the more salt-resistant barley. Eventually barley yields declined, agriculture foundered, villages and cities were abandoned, and the famous civilization came to an end. Today there are still thin, white layers of salt covering the once fertile soil. Salination is frequently a problem in irrigation projects and expensive remedies are employed in the United States and the Middle East to avoid permanent degradation of the land.

Barren land may result from poor management even in areas of rainfall sufficient to allow the development of forest. Many tropical clay soils, in contrast to temperate clay soils, contain very little silica in their structure; but they contain tiny crystals of hydroxides of aluminium and iron that hinder the retention of chemical nutrients in the soil spaces. As warm rain percolates through soil cleared for cultivation the nutrients are washed away and the soil is quickly exhausted. The proportion of iron and aluminium hydroxides increases and eventually a rock-hard crust of laterite is formed. Buried laterite, the result of events in the geological past, does not cause much trouble; but erosion can bring it to the surface where it becomes so hard that it can be quarried. Many of the older houses in Freetown in Sierra Leone are built from laterite blocks, and laterite forms the main structure of the Khmer ruins in south-east Asia. Laterization is a growing problem in tropical countries whenever forests are felled. The long-term future of land once covered by the magnificent Brazilian forests is bleak. Cleared strips along the main highways receive less rainfall than the adjacent forest and at present rates of forest clearance there may be nothing left in 35 years but unproductive, severely laterized land which is unlikely to support viable cultivation.

The mistakes of the recent past, let alone of ancient civilizations, are unlikely to influence policies in over-populated, impoverished, and under-developed countries where the advantages of immediate returns from cash crops outweigh the fears of long-term land degradation. Available land for cultivation is limited and over-exploitation in the interests of quick returns will soon make things much worse. There is a problem now, but unless well thought-out, well-financed policies prevail vast areas of land will soon be unfit for human use. Egypt has embarked on an ambitious programme to make the desert bloom and double the area of arable land. Silt accumulated in Lake Nasser behind the Aswan High Dam will be used to improve the fertility of desert soils, which will be irrigated by a canal network fed from the lake. Ironically, Sudan's proposed Jonglei Canal will divert the Nile and prevent it spilling over into the Sudd and hence reduce the area available for cultivation and grazing. Water is needed in northern Sudan and it is felt that too much is lost from the Nile as a result of spillage into the Sudd and consequent evaporation.

It is certainly possible to reclaim land that has been degraded by mistakes in management: the former North American dustbowl and the ruined lands of Kazakhstan in the Soviet Union have been in part restored, but at enormous cost. Needless to say few countries can afford massive expenditure on land reclamation and we are left with the conclusion that in many parts of the world productive land will continue to be lost.

## The limits to exponential growth

The eating of snails by thrushes described earlier in this chapter should indicate to us that the rate of consumption of a resource will necessarily fall off rapidly as the resource becomes scarce. If there are no alternatives the consumer is faced with a problem, and this is precisely the predicament beginning to face man. Nowhere is this predicament more clear than in present trends in oil consumption.

All populations have the potential for exponential growth in size and when this occurs there is a tendency for exponential consumption of resources. This tendency is to a large extent resisted by the environment and the result is a balance between what organisms require and what the environment can provide. In the next chapter we examine what limits exponential growth and consumption in nature.

# 3 How populations are regulated

India has for long attracted attention as a nation where the pressures of a large human population are likely to be felt sooner than anywhere else. Indeed the very mention of India conjures up an image of teeming millions of people and appalling poverty. Every year we hear of famine or the possibility of famine and we are urged to give generously to save people from starving. Unexpected drought and the failure of the rains are usually blamed, but drought seems to occur with such monotonous regularity that we could be forgiven if we suggest that Indians should stop calling the predictable the unexpected.

At the time of the last official census in 1971 there were about 548 million people in India and the population was increasing at an annual rate of 2·5 per cent, which means that each year at least 13 million people are being added. The present (1978) population must be more than 640 million and if the current rate of increase is continued there will be over 1000 million Indians by the turn of the century. Millions of Indians have adopted birth-control or accepted sterilization, but because India's population is so very large this has had virtually no impact on population growth, indeed the present rate of increase is actually higher than the annual growth rate of 2·2 per cent operating in the 1950s. In 1920 births and deaths were balanced at about 49 per thousand but since then the application of medical knowledge has reduced the annual death rate to about 14 per thousand while the birth rate has been only slightly reduced and indeed at times has been higher than 49 per thousand. Medical and social workers in India are often well-satisfied with local successes of schemes for limiting population growth but the total situation is fast becoming unmanageable. An additional and also (it is claimed) unexpected problem is that conflicting religious interests generate suspicion of the motives of those wishing to limit population growth by the introduction of birth-control methods. Thus the Hindus, who constitute over eighty per cent of the population, believe that the Muslims are increasing in numbers more rapidly than themselves and hence are suspicious of government intentions. An emotive factor in the accumulation of evidence by the Janata Party for a court case against the former Prime Minister of India, Mrs Indira Ghandi, was that sterilization was enforced. Birth-control, seemingly such a good idea, frequently runs into difficulties of this kind. Almost everywhere in the world people of different ethnic groups or religious persuasions are suspicious of others, and this suspicion is nowhere more intense than over proposals that might threaten population size and the right of individuals to have as many children as they want.

The present rate of increase of India's population cannot continue for much longer. Soon the environment will begin to exert pressures that will stabilize population growth; almost certainly it will be famine which will cause a fall in the rate of growth, but disease must also be considered as a likely alternative, and the possibility of the two acting together is considerable. India cannot expect to go on receiving massive gifts of food from other countries – food already needed by the donor countries – and its population will therefore be limited by its own capacity to produce food. It may seem a gloomy thought, but every time you give money to a charitable organization to relieve famine in India you are making a small contribution to postponing what is inevitable.

India's position is by no means unique. Much the same applies in many other countries, but because the population of India is so very large and because a crisis seems imminent we shall probably gain from it an insight as to what will eventually happen throughout the world. The world population is now over 4000 million. The President of the World Bank predicts that 'if urgent measures were implemented' it could stabilize at 8000 million by the year 2070. If no such measures are taken it could reach 11,000 million and then stabilize under conditions of poverty, stress, hunger, crowding, frustration, and disease. The urgent measures suggested include improving the education, status, and rights of women throughout the world which, it is thought, would act as a major factor leading to the required massive decline in child-bearing.

Our interest in this chapter is to try and understand how populations of most organisms exist in a state of balance with the resources of the environment. By doing this we shall perhaps be able to see our own position in ecological perspective.

## Carrying capacity

The resources of the environment can be considered mainly in terms of the availability of energy and nutrients. Green plants capture radiant energy from the sun and by the process of photosynthesis* convert it to stored chemical energy. All other organisms depend directly or indirectly on green plants for their food. It would therefore seem that ultimately populations of animals are limited by the ability of plants to convert the sun's energy into food, but this is a simplification and we must define the resources of the environment rather more broadly.

The individual members of all plant and animal populations interact with each other in a variety of ways. They may compete for the same sources of energy or nutrients or for space to such an extent that an individual's chances of survival depend on how many other individuals are present at the same time. The ability to escape predators and parasites, to resist disease, to find cover and shelter, and to avoid bad weather are all potentially capable of determining the limits of population growth. Because there are so many factors involved and because not all of them seem directly concerned with the acquisition of energy and nutrients, the term 'carrying capacity' is often used to mean the ability of the environment to support a population. This carrying capacity must be understood in terms of a particular population and must include everything that affects the size of that population.

If a population is growing exponentially it will, under normal circumstances, quickly reach the carrying capacity of the environment. As the carrying capacity is approached the rate of population growth will slow down and eventually the death rate will be equal to the rate of recruitment of new individuals into the population. When this occurs no further growth can take place and the population is regulated at a fixed level. A regulated population is one that returns to a balanced level after departure from that level. There are several possible ways in which exponential growth is stopped and three of these are shown graphically in Figs. 7–9. It is, however, important to remember that the events shown in these graphs can rarely be observed directly, as in most real populations death and recruitment are balanced over long periods of time and nothing but small fluctuations in population size can be detected. In each of the three

*Photosynthesis is explained in Chapter 5.

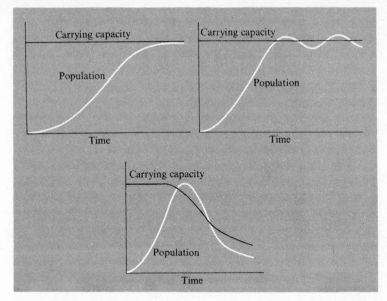

*Fig. 7:* Increase in population size and stability resulting from pressures exerted by the carrying capacity of the environment.

*Fig. 8:* Increase in population size and instability resulting from pressures exerted by the carrying capacity of the environment.

*Fig. 9:* Increase and decrease in population size resulting when a population reduces the carrying capacity of the environment.

graphs the population is represented as the number of individuals, but the carrying capacity cannot be defined so precisely: it is simply a measure of the ability of the environment to support a population of a species over a period of time and as such cannot be expressed in any particular units. In the simplest sense it can be thought of as a measure of the availability of food.

In Fig. 7 the population initially grows quickly, but as the carrying capacity is reached the rate of growth slows down and stops. The population is then at equilibrium with the carrying capacity of the environment. This is theoretically what happens every time a species of plant or animal extends its range and establishes a new population. An introduction to a new area, such as the pheasants on Protection Island (Fig. 6) approximates this pattern. Much the same happens in Fig. 8, except that when the population reaches the carrying capacity it fluctuates within limits above and below what the environment is offering. This is probably the usual situation although actual figures are rarely available. Populations of small birds and mammals increase in years favourable for feeding and decline in adverse years, which means population size fluctuates around the long-term carrying capacity. This may generate a certain amount of instability which could lead either to the situation shown in Fig. 7 or to that in Fig. 9. The same exponential growth occurs in Fig. 9, but this time the carrying capacity is reduced by pressures exerted by the population, and there is subsequently a fall in both the carrying capacity and in the population. This is equivalent to saying that the population is depleting its own resources. The human population in the Sahel has followed the theoretical pattern shown in Fig. 9. Carrying capacity was reduced as increasing numbers of people and their livestock damaged the environment which is now unlikely to recover. Population decline has been arrested only by the injection of food hand-outs from the rich countries; without such help there would have been a dramatic and irreversible population crash.

But most animal populations, even the big mammals of East Africa with all their admirers in the rich countries, do not receive help on the same scale. In the mid-nineteenth century wooded savanna teeming with mammals covered Kenya from Mombasa on the coast inland to Mount Kenya. The railway, built in 1896, opened the country to traders, settlers, hunters, and (later) to souvenir-hungry tourists. The vegetation thinned and the mammals decreased. In 1948 Tsavo National Park was established, providing 20,800 km$^2$ of countryside within which elephants and other species would supposedly be protected by law. But even this vast area proved too small, and the elephants constantly left the confines of the Park to seek food elsewhere. Those that did were usually killed. A full-grown elephant needs about 180 kg of food a day, most of it in the form of 'browse' from trees and bushes. By 1967 there were two elephants to every square kilometre of the Park, obviously more than the environment could support. What was once woodland became grassland and is now fast becoming desert. Starving elephants even chew dry tree trunks and suffer intestinal blockages; meanwhile their habitat collapses around them. The elephants began dying in numbers in the 1960s, and there are now perhaps 5000–10,000 left in Tsavo – no one knows the exact numbers. But those that remain live in an impoverished, barren land that cannot possibly continue to support them. The Tsavo elephants demonstrate vividly the events graphed in Fig. 9 when the carrying capacity is reduced by the population it is supporting. The graph could be continued in a number of different ways. Thus the carrying capacity, released from pressures exerted by the population, could rise again to its former level, or both the population and the carrying capacity could continue at a lower level than before. In the case of the Tsavo elephants it is unlikely that the land will recover and the elephant population is probably facing total destruction: there is simply not enough space.

If we were able to measure the size of a population and the carrying

*Plate 5:* Elephants in Tsavo Park, Kenya, have destroyed woodland and created grassland, which is fast becoming desert.

capacity offered by its environment over a long period of time we should nearly always find that although both would fluctuate a little the long-term pattern would be of equilibrium. This is the usual state of affairs, despite the ability of populations to expand exponentially. But which of the three graphs most accurately fits the future situation for India? Almost certainly it is Fig. 9.

Because of current social and political attitudes it is probably inaccurate to speak of 'the population of India' and 'the carrying capacity of India'. Increasingly the international community is creating a world where the only meaningful assessment of carrying capacity for people is on a global scale. One country's population and food problems must be seen as the world's problems. The alternative is the horrifying possibility of international competition for food.

## Limiting factors

Everything that affects the size of a population is part of the carrying capacity of the environment, but there are usually just one or two factors that play a really important role in population regulation. When they can be identified these are called 'limiting factors', and as the carrying capacity of the environment is approached it is these that exert the most severe restraints on further population growth. The concept of limiting factors as generally used and understood is a combination of several different principles relating organisms to their environment. In the mid-nineteenth century, Liebig realized that 'the growth of a plant is dependent on the amount of foodstuff which is presented to it in minimum quantity', a principle now known as Liebig's 'law' of the minimum. In essence this means that the photosynthetic activity and rate of growth of a plant are limited not by abundant materials but by rare, or trace, elements necessary in minute amounts. Liebig's law has been extended to include the effects of temperature and light intensity as well as of nutrients on the rate of photosynthesis.

The historical development of the concept of limiting factors started by a consideration of what limits plant growth, but has been widened to include everything that affects growth and reproduction in populations of plants and animals. Limiting factors are most easily understood when resources such as food are considered, because the relationship between resource supply and population density is often obvious. It is less easy to appreciate how an environmental variable such as temperature acts as a limiting factor. Variation in temperature within the limits of tolerance of a plant affects the rate of photosynthesis and of growth; low temperatures slow the process, and at high temperatures water loss and wilting interfere with it. Extremely low or high temperatures may be lethal. Temperatures may also affect both the abundance of a food resource and its availability to an animal. Cold weather in early summer inhibits the growth of populations of aphids that form the aerial plankton on which swifts feed; even if aphids are abundant, cold weather stops them flying which makes them unavailable to swifts.

Traditionally water has been the limiting factor for people and livestock in arid regions of the world. The sinking of deep bore-holes, often financed by foreign aid, has made water more available. In places water is no longer in short supply and livestock have been allowed to increase in number up to and above the carrying capacity of the environment. The availability of grazing is now the limiting factor.

The population density of hole-nesting birds is frequently limited by the availability of suitable nesting sites and populations may be held well below the limits set by food. But placing nest-boxes in gardens and conifer plantations encourages a far higher nesting density of blue tits and great tits than would otherwise occur. Gardens and conifer plantations rarely support enough insects for the extra adults to feed the nestlings and in most years nestling mortality is high. This parallels the situation in arid regions – remove one limiting factor and another takes its place.

## Competition

In view of the fact that the carrying capacity of an environment is limited, it seems likely that there is competition among individuals for the resources. Competition in this sense need not necessarily imply open conflict between individuals. Even among mammals and birds it is unusual to see fights to the death, but threatening behaviour is commonly observed, as for instance among robins at a feeding table in winter. Robins and other birds frequently displace one another from a source of food, one bird quickly responding to the threatening behaviour of another and moving aside. We may suppose that repeated displacement from a food source will lead to deprivation and, at times of food shortage, to death. Similarly if two seedlings start growing in a patch of soil where there is really only space for one, the stronger and more vigorous will eventually displace and cause the death of the weaker.

By competition we must therefore envisage a situation in which two or more individuals are consuming resources of limited availability. Those best able to exploit the limited resources will survive; the remainder will perish. The intensity of competition varies with the density of the population. The more individuals the more likely that the carrying capacity of the environment will be subject to stress. Much of the mortality that occurs varies with the density of the population and ecologists often speak of death rates as being density-dependent, which means that the probability of an individual's death depends on the density of the population in which it occurs. The availability of food is likely to affect survival and death density-dependently, but unusual weather, such as drought or flooding, may affect all members of the population independently of density. The ecological literature of the past 20–30 years is dominated by discussions as to whether the main source of mortality in populations occurs density-dependently or density-independently. Much seems to depend on the kind of situation being analysed: ecologists working on populations of woodland birds are impressed by density-dependent factors, while those working on insects in dry regions are more inclined to invoke density-independent factors.

## What is an individual?

We have already seen that individual members of a population compete for resources, but we must here pause for a small digression and ask exactly what is meant by an individual. The question is partly philosophical but also biological. We have no hesitation in saying that the robins in the garden are different individuals, and we would probably say the same about honeybees. But honeybees, like ants, bumblebees, and the large yellow and black wasps, are social insects living in nests in which only the queen reproduces. Most of the bees we see are non-reproducing female workers that never mate and so contribute nothing to the next generation. Should they be regarded as distinct individuals or are they simply 'part' of the queen, the part that forages and brings food back to the nest to feed the larvae? Worker honeybees are produced by a sexual process and because of this they are all slightly different from one another: in this sense they are different individuals. But in ecological terms it is probably more useful to think of competition between colonies of honeybees than between workers that are all part of the same colony. There is in effect a division of labour the function of which must be seen in terms of survival of the colony rather than the survival of each worker.

Every organism produced by a sexual process inherits characteristics from each parent. Sex cells (eggs and sperm) are produced by segregation and recombination of the hereditary material and as a result they are all different from one another and give rise to genetically distinct individuals. This offers a useful biological basis for individuality, one that implies genetic distinctiveness. But what about the many organisms, animals as well as plants, that reproduce without a sexual process?

Many species of grass put out long horizontal stems – called stolons if they are above ground and rhizomes if below ground – along which new grass plants develop. In this way a thick sward of grass can arise from a few scattered plants. Much of the grass is produced by asexual vegetative growth. Thus there may be thousands of grass plants in a field but only a few genetic individuals. Blackberry bushes also spread by vegetative reproduction: long woody stems arch over and root soon after touching the soil. Once the roots grow the linking stem may break and both new and old plants survive. One blackberry bush or two? Genetically they are identical but physically they are separate. The English elm produces abundant root suckers often at some distance from the parent tree. These suckers can grow into new trees physically separate from the parent tree. Unlike the wych elm, the English elm rarely (perhaps never) produces fertile seed. A hedgerow of elms is usually made up of the well-grown suckers of one tree. It is almost as though the English elm has penetrated the entire Midlands and south of England by an underground network, putting up trees at intervals wherever there is a vacant space. We only see what is above the ground and therefore tend to consider them as separate trees but their roots are (or have been) continuous. The English elm is probably a hybrid between the native wych elm and one of the continental species introduced hundreds of years ago. Sometimes artificially produced hybrids between species fail to produce viable seed and there is thus the intriguing possibility that all English elms are the same genetic individual.

Familiar pests such as aphids present similar problems of individuality. The first aphids in spring hatch from over-wintered eggs produced by fertilized females the previous autumn. Then throughout the summer there follow a series of generations during which females give birth to live female offspring produced without any sexual process, a method of reproduction called parthenogenesis. A single female can give rise to millions of aphids all genetically identical to herself. Males appear in autumn and mate with females which lay the over-wintering eggs. The huge clusters of black aphids which infest our bean plants in summer are produced by parthenogenesis and from one point of view are all the same individual. When we attempt to destroy them we kill 'parts' of a gigantic super-aphid and rarely succeed in destroying the entire individual, which can of course go on reproducing.

Aphids, elms, and certain other organisms exist as two sorts of individual: separate physical individuals and genetic individuals consisting of large numbers of 'parts' or 'bits'. It follows from this that the size of their populations must be viewed in two ways and that competition between individuals presents special problems of analysis; there are also problems of evolution and adaptation, a topic we shall take up in Chapter 6.

## Density-dependence and the stability of populations

Let us see whether changes in density in populations caused by reproduction have any effect on population stability. Plants and animals living in temperate climates tend to reproduce only at certain and predictable times of the year. Everyone living in a temperate region is familiar with the annual cycle of flowering and fruiting, birds breeding in the spring, and insects becoming common in the summer. It is a little more complicated in the tropics, reproduction in many species occurring throughout the year, but in almost all species there are seasonal peaks of reproductive activity.

Seasonal reproduction results in large numbers of young plants and animals appearing suddenly in a population. Most of them fail to survive for more than a short time. Few of the seeds produced by plants result in mature individuals which can themselves reproduce, and most of the eggs and young produced by animals do not result in breeding adults. The death rate of young individuals is extremely high, often over ninety-nine per cent. Consider for example a female butterfly that lays two hundred eggs. If the population of this butterfly is to remain stable, then on average only one male and one female capable of breeding must survive until the next generation, and the remaining 198 must die as eggs, caterpillars, or pupae. Losses occur at all stages of development, but investigations into the factors causing death show that it is the young individuals that suffer most severely, and that once a butterfly, bird, tree, or indeed any other organism has become a reproducing adult its chances of subsequent survival are much improved. This is something of a generalization and although widely applicable to the more familiar plants and animals around us, it should not be understood to apply necessarily to all organisms. Indeed for some like viruses and yeast cells the concept of an adult has little meaning, as reproduction occurs quickly and there is no young stage that can be clearly differentiated.

The effects of seasonal reproduction on a population may be seen in the great tit, a common woodland bird in Europe. Great tits normally breed in holes in trees and by providing nesting boxes it is possible to entice all the birds in an area to breed in them, which makes much easier the gathering of information on the number of eggs laid and young produced. In southern England great tits lay 10–12 eggs in April or May, or earlier if it is warm, and for reasons not fully understood more eggs are laid in early than in late years. Most of the young of a brood hatch on the same day and are fed by their parents for just over two weeks on caterpillars of certain moths which in May and June are abundant on the leaves of trees such as oak and elm. The breeding success is high and the majority of eggs laid produce flying birds in the early summer. These young birds feed on a variety of insects but in most years become short of food by July and August because many of the kinds of insects eaten by great tits are unavailable.

When the young leave the nests there is something of a population explosion in the woods. Each pair has produced 10–12 young, so that if an area of woodland supports 20 breeding pairs there is a total population of 240–280 soon after the young fledge. Yet by the beginning of the next winter the population is back to about 40 birds, many of them the same individuals that bred in the wood earlier in the year, some newcomers from elsewhere, and a few birds born in the wood. In the summer the tits disperse into other areas, but many disappear, having presumably died. The causes of death are not easy to determine as in summer the woods are overgrown and the birds silent and inconspicuous. But whatever happens it is clear that the sudden increase in late May and June does not lead to a large population in the winter, and by the time the next breeding season comes around the population is back to normal. There are of course fluctuations in the size of the breeding population from year to year but they are nowhere near as great as would be possible considering the number of young added each year.

Fig. 10 shows the seasonal events in a population of great tits breeding in a small wood near Oxford. Here it was possible to estimate the number of birds present at four different times of the year. The

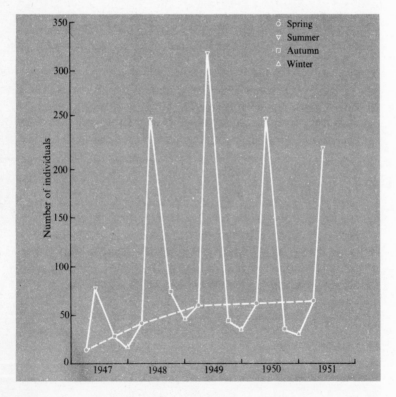

*Fig. 10:* Seasonal and annual changes in the size of a population of great tits in a wood near Oxford. (From D. Lack, 1954)

birds were counted in winter when they are easy to see, and again when they bred in nesting boxes in spring. The number of young produced was known which gave an early summer estimate, but another census was not feasible until autumn when the leaves had fallen from the trees. As shown in Fig. 10 the breeding population was low in 1947, possibly because of the severity of the previous winter, but subsequently remained relatively stable. The rise from 1947 to 1949 might have been caused by more and more birds accepting the boxes for nesting. As shown in Fig. 10 the population increased five- or six-fold each summer when the young left their nests but by autumn numbers were down again. Not all of the young tits necessarily died, but the fact that the majority did not remain in the wood suggests that the carrying capacity of the local environment had been reached. Every year there was a massive increase which had no long-term effect on the great tit population in the wood.

The sequence of events described for the great tit occurs in most other populations: after reproduction there is an enormous temporary increase but the population quickly falls to its original level. Things are quite different with man. In man breeding occurs throughout the year (although in some societies there are slightly more births at some seasons than others) and deaths too are rather evenly distributed in time. Moreover modern medicine and agriculture have made possible the survival of babies and children who would otherwise have died and there is nothing like the death rate of immature individuals which occurs in virtually all other populations of plants and animals.

## Heron populations in Britain and North America

There is at present no complete picture of how a population is regulated in any species of plant or animal living in a natural or semi-natural state. Even in experimental populations of insects and micro-organisms in the laboratory where the environment has been deliberately simplified and controlled there is still much to be learnt as to exactly how population growth is stabilized at a particular level. But rather than discussing experiments on unfamiliar organisms we shall in this section consider birds of well-known species, describing in detail what is known of the population dynamics of herons. Much is known about herons but the picture that will emerge is by no means complete and we shall be left with some unanswered questions.

The two species under consideration, the European grey heron and the North American great blue heron, replace each other on either side of the Atlantic. They differ in a number of minor details of structure and plumage but they are so similar that we could with justification treat them as the same. Both species are large and conspicuous and are known to naturalists and especially to anglers who may consider them a danger to fishing interests.

Herons feed chiefly on fish which they grab from shallow water with their beaks. Both freshwater and sea fish are eaten but herons are especially associated with freshwater, and relatively few depend on sea fish. They also eat small mammals, newts, young birds, crabs, shrimps, and a variety of aquatic insects, but they are essentially fish-eaters and almost everything about their way of life is oriented towards catching fish. Since herons cannot dive for food and rarely swim, the water they fish must be shallow enough for wading and clear enough for the fish to be visible.

Herons breed in colonies which sometimes contain a hundred or more bulky nests built high in trees; many colonies are much smaller, and a few pairs breed in isolation. The same nesting sites are used year after year. In areas where there are few trees they may nest on the ground or among rocks, but this is unusual. The largest colonies occur in regions where there is an abundance of shallow water suitable for fishing, and small colonies and isolated nests are a feature of countryside where the streams are fast-flowing and where fishing is more difficult. Heron colonies are spaced in a way that maximizes the use of fishing resources, which strongly suggests that during the breeding season there is a premium on good fishing grounds. This kind of spacing is common in plants and animals. Its existence has led ecologists to assume that bird populations in particular are limited in size by the availability of food. The notion of availability is especially important in a bird like the heron: deep water with plenty of fish cannot be exploited and what is required is shallow water where fishing is possible. This difference between abundance and availability of food comes up repeatedly when we consider whether food is important in regulating the size of populations.

Herons breed early in the year, the dates of egg-laying depending on climate and latitude. In southern England clutches of eggs are started in early March with a peak in the third week of the month; in warm springs laying may be earlier and in cold springs later than usual, but most clutches are started by the middle of April in almost every year. If the eggs are lost (crows sometimes take them from an unguarded nest) the female will lay again and this accounts for occasional late nests. The eggs are laid at intervals of about two days and incubation begins on the second or third egg which means that the young hatch at two-day intervals. The size of the clutch varies between two and five, rarely six; in southern England four is the average. Herons do not usually breed until they are two years old.

In 1928 amateur bird-watchers in Britain combined to count as many nests as possible of the grey heron. This census has been repeated every year and although some nests have obviously been missed there is probably a more complete picture of long-term population trends in the grey heron than for any other species of bird. The nests are counted late in the spring when most pairs would be expected to have young. At this time it is possible to judge from the ground (many heron nests are 20 metres or more up in trees) which nests are occupied and which are not. Occupied nests have fresh green twigs around the rim and there is clearly visible 'white-wash' caused by the liquid excreta of the young.

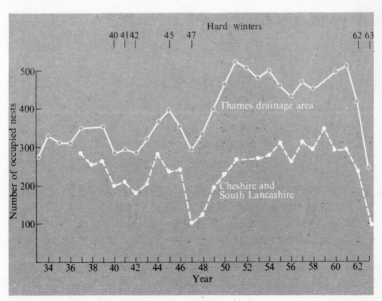

*Fig. 11:* Number of breeding pairs of grey herons in two areas of England. (From D. Lack, 1966)

The census is particularly accurate for two areas of England: the Thames Valley, and Cheshire and south Lancashire. The number of nests in these two areas during about thirty consecutive years is plotted in Fig. 11. Admittedly at first sight the graph suggests large fluctuations in the number of nests from year to year, but the fluctuations shown are in fact small compared with what is potentially possible. Thus in many years a pair of herons successfully produces about three young which means that if there are 400 pairs breeding in the Thames Valley a further 1200 are added after the breeding season, giving a total of 2000. As shown in Fig. 11 there is never an increase in the breeding population that even approaches this figure. It follows therefore that between breeding seasons many herons die or disappear from the area and the long-term result is that although there are fluctuations in the population they are small compared with what is theoretically possible. Many of the herons breeding in Britain remain all winter. In hard winters (marked in Fig. 11) it is likely that food becomes scarce and that many herons die of starvation. A bird dependent on food from shallow water is especially liable to suffer when the water freezes. There is usually a smaller breeding population after each hard winter but it does not take many years for the population to reach its original level. Fig. 11 shows that the populations in the Thames Valley and in Cheshire and south Lancashire fluctuate in parallel: a poor year in one area is also poor in the other, which further emphasizes the effects of hard winters.

Ornithologists have for years been placing numbered metal rings on the legs of birds in order to find out about migration and death rates. These rings request anyone finding a bird to report it to a central ringing office. The frequency of recovery of ringed birds depends on the species. There are relatively few recoveries of small species or species which migrate to remote areas where the chances of reporting a recovery are low. But for large birds like herons the recovery rate is high. Putting rings on the legs of nestling herons in not easy as it usually involves climbing high into the tree tops, an exercise which few can perform with competence, but even so many nestling herons have been ringed and there are a sufficient number of recoveries to obtain an estimate of the death rate.

Fig. 12 shows the distribution of deaths by age and month of the year of young great blue herons after they have left the nest, and of adults (birds at least a year old) by month of the year only. As can be seen

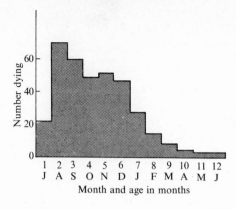

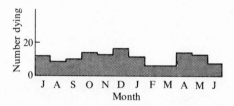

*Fig. 12:* Death rates of North American great blue herons. The upper histogram shows the number of ringed birds recovered each month during their first year of life. The lower histogram shows the number of adult birds recovered each month.

many young birds die during the summer soon after they leave the nest and the death rate remains high until the following February when it falls sharply. Many herons die when only a few months old, possibly because of inexperience in catching fish. There are no obvious seasonal changes in the likelihood of death for adult birds. Evidently the critical time for herons is in their first six months; thereafter their chances of survival are much improved. The figures are even more dramatic if deaths are plotted by age in years as in Fig. 13. The high death rate in the first year falls steeply in the second year and by the time the birds are in their third year and ready to breed it has fallen even further. From then on the death rate is low; rather few birds seem to live for more than eight years, but as can be seen in Fig. 13, a bird can live until it is twenty-one years old.

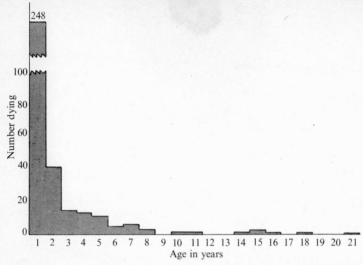

*Fig. 13:* Death rate of North American great blue herons in relation to age.

These statistics for great blue herons can be summarized in a somewhat different way and compared with those for the grey heron, as shown in Table 3. The results for the two species are almost identical which perhaps is to be expected since they are so similar, but we must bear in mind that the figures for the great blue heron refer to the whole of North America while those for the grey heron are for birds born in Britain, relatively a very small area. Both species experience a high death rate in their first year and the average annual death rate after the first year is about thirty per cent. This means that on leaving the nest a heron can on average expect to live no more than eighteen months, but if it survives the first year it can expect a further three years of life. Thus most herons die young, many before they breed, but those that survive for more than a year can expect to breed several times, and a few may live a long time. The high death rates of young are typical of virtually all other species of birds that have been studied in this way. We ourselves are used to associating death with old age and these results may come as something of a shock: it is perhaps a sobering thought that most of the birds born in your garden will die when less than a year old and before they have a chance of nesting.

As mentioned earlier the eggs of herons are laid at intervals of two

*Table 3* Comparison of death rates and chances of life in the great blue heron and the grey heron

|  | Great blue heron (North America) | Grey heron (Britain) |
|---|---|---|
| Number of ringing recoveries used | 349 | 195 |
| Death rate in first year (per cent) | 71 | 69 |
| Average annual death rate after first year (per cent) | 29 | 31 |
| Expectation of life after leaving the nest (years) | 1·5 | 1·5 |
| Expectation of further life at beginning of second year (years) | 2·9 | 2·7 |

days; incubation begins when the second or third egg has been laid, and no development takes place inside the eggs until the start of incubation. The first two or three nestlings hatch on the same day, and then there are usually intervals of about two days between subsequent hatchings. Since the young nestlings grow rapidly there is a disparity in size between those hatched first and those hatched later. Quite often the last to hatch is extremely small compared with its older brothers and sisters. This sequence of hatching does not occur in all species of birds; it is associated especially with large predatory birds and is less common in small song birds (like the great tit) whose young tend to hatch together. We shall return to the significance of staggered hatching when we have discussed the food of nestling herons.

The nestlings are fed by both parents who catch fish in waters far removed from the colony and return at intervals with their stomachs full of food. After being caught the fish are swallowed head first. On delivery back at the nest they are regurgitated tail first, turned round and fed to the nestlings head first, which makes it much easier for the young birds to swallow spiny fish like perch. If an adult has been feeding for a long time the food brought back is partly digested but if it returns quickly the food items are more or less intact. Like many other fish-eating birds the nestlings regurgitate food when disturbed by an intruder – presumably to divert the attention of potential predators – and if you climb to a nest containing nestlings you will be greeted with a hail of spat-out fish. This behaviour may be exploited as a means of finding out exactly what they eat. Many of the food items are readily identifiable and can be replaced in the nest to be eaten again by the nestlings.

*Plate 6:* The author climbing to a heron's nest high in an oak tree at High Halstow in 1949.

The food of nestling herons has been studied at three colonies in the Thames Valley: Wytham, near Oxford; Buscot, about 32 km west of Wytham; and High Halstow, about 150 km east in the Thames estuary. Table 4 lists the food items found in samples from nestlings at each of these three colonies. The list contains fish and other animals familiar to anglers, some of the commoner species of fish being the same as those sought by anglers in the Thames Valley. There are striking differences in the food brought to nestlings at the three colonies. At Wytham and Buscot coarse fish are most important; substantial numbers of trout are taken at Buscot but not elsewhere, while at High Halstow eels, sticklebacks, and shrimps form the staple diet. These differences are associated with the different environments available to the hunting adults. At Wytham and Buscot they feed mainly along the banks of rivers; around Buscot the rivers are smaller and faster-flowing than around Wytham which presumably explains the higher frequency of trout at Buscot. At High Halstow the birds feed in drainage ditches and pools on reclaimed pasture, and to a small extent in the brackish water of the estuary itself. Many of the ditches also contain slightly brackish water and this explains the presence of shrimps in the samples. Table 4 shows that mammals and birds are occasionally taken by herons, but apart from the water vole they are unusual, although the capture of an adult stoat and its subsequent feeding to a nestling deserves comment as the stoat is itself a vicious predator. The insects listed seem to be taken incidentally while the herons are waiting for larger prey. Most of the dragonflies are newly emerged and too young to fly properly, and the beetles are large aquatic species. The absence of frogs in these samples is striking and it might be that during the period of investigation frogs were scarcer than usual in the Thames Valley. The adult herons do not of course catch all the fish they encounter. Specimens more than 25 cm long are too big, and most of the roach, perch, pike and similar species taken are between 11 and 20 cm in length, smaller than those regarded as interesting by the average angler. Apart from shrimps and sticklebacks, prey items less than 10 cm in length are usually avoided, even though they must be abundant where herons feed. Sticklebacks and shrimps are possibly taken when larger items are unavailable.

Table 4 contains an impressive list of animals and includes many of the familiar species you would expect to see in and around the River Thames. It suggests that the heron is an opportunist while feeding, seizing anything it can reasonably manage, but concentrating on medium-sized fish whenever possible. Because of its method of feeding we should expect the availability of food to vary from day to day, with the season of the year, and from year to year, something anglers expect and find whenever they try their luck with a rod and line. We should further expect that changes in the availability of food will affect the rate of growth and survival of the nestlings. It seems that late in the season (June and July), when the nestlings are beginning to leave the colony, food becomes less available because of growth of plants in the water and along the edges of rivers and drainage ditches. At High Halstow the water in the ditches is virtually unexploitable late in the season because of a dense growth of aquatic plants on the surface. Dry weather is better for fishing than wet, possibly because after rain the water becomes muddy which makes fishing more difficult. These and other factors determine the ability of the adults to feed their young successfully.

An idea of changes in feeding conditions can be obtained by weighing the nestlings at intervals to see how they grow. Changes in body weight are often used as an index of food and nutrition in birds and mammals, including human babies. Fig. 14 shows the weights of

*Table 4* The food of nestling grey herons at three colonies in the Thames Valley

| | Wytham 1952–7 | Buscot 1953–7 | High Halstow 1953–6 |
|---|---|---|---|
| *Number of food samples examined* | 313 | 131 | 160 |
| **Fish** | | | |
| Trout, *Salmo trutta* | 2 | 45 | — |
| Pike, *Esox lucius* | 14 | 3 | — |
| Carp, *Cyprinus carpio* | 2 | 5 | — |
| Gudgeon, *Gobio gobio* | 83 | 12 | 3 |
| Tench, *Tinca tinca* | 7 | 4 | — |
| Roach, *Rutilus rutilus* | 539 | 122 | — |
| Rudd, *Scardinius erythrophthalmus* | — | — | 56 |
| Bleak, *Alburnus alburnus* | 149 | — | — |
| Minnow, *Phoxinus phoxinus* | 158 | 195 | — |
| Eel, *Anguilla anguilla* | 57 | 14 | 203 |
| Perch, *Perca fluviatilis* | 94 | 10 | 1 |
| Ruffe, *Acerina cernua* | 13 | 1 | — |
| Bullhead, *Cottus gobio* | 13 | 11 | — |
| Stone Loach, *Nemacheilus barbatula* | — | 68 | — |
| Goby, *Gobius minutus* | — | — | 11 |
| Sticklebacks, *Gasterosteus aculeatus* and *Pygosteus pungitius* | 285 | 162 | 1031 |
| Great pipe fish, *Syngnathus acus* | — | — | 5 |
| Flounder, *Pleuronectes flesus* | — | — | 1 |
| Unidentified | — | — | 1 |
| Total fish | 1416 | 652 | 1312 |
| **Mammals** | | | |
| Water vole, *Arvicola amphibius* | 42 | 16 | 13 |
| Field vole, *Microtus agrestis* | 1 | 2 | 2 |
| Mole, *Talpa europaea* | 4 | 3 | 1 |
| Water shrew, *Neomys fodiens* | 1 | 2 | 2 |
| Common shrew, *Sorex araneus* | 5 | — | — |
| Pygmy shrew, *Sorex minutus* | — | 1 | — |
| Stoat, *Mustela erminea* | — | — | 1 |
| Total mammals | 53 | 24 | 19 |

|  | *Wytham* 1952–7 | *Buscot* 1953–7 | *High Halstow* 1953–6 |
|---|---|---|---|
| **Birds** | | | |
| Coot, *Fulica atra* (chick) | — | 1 | — |
| Moorhen, *Gallinula chloropus* (chick) | — | — | 3 |
| Lapwing, *Vanellus vanellus* (chick) | — | — | 1 |
| Total birds | — | 1 | 4 |
| **Amphibians** | | | |
| Newt, *Triturus* sp. | — | 6 | — |
| **Crustaceans** | | | |
| Shrimp, *Palaemonetes varians* | — | — | 1059 |
| Shrimp, *Crangon vulgaris* | — | — | 38 |
| Shrimp, *Gammarus* sp. | — | — | 5 |
| Crab, *Carcinus maenas* | — | — | 2 |
| Crayfish, *Astacus fluviatilis* | — | 2 | — |
| Total crustaceans | — | 2 | 1104 |
| **Insects** | | | |
| Dragonflies, 6 species | 49 | 1 | 4 |
| Beetles, 2 species | 21 | 3 | 40 |
| Bugs, 3 species | 5 | 1 | 6 |
| Scorpion flies, 1 species | 1 | — | — |
| Caddis flies, 1 species | 1 | — | — |
| Flies, 2 species | 1 | 1 | — |
| Total insects | 78 | 6 | 50 |
| Total items | 1547 | 691 | 2484 |

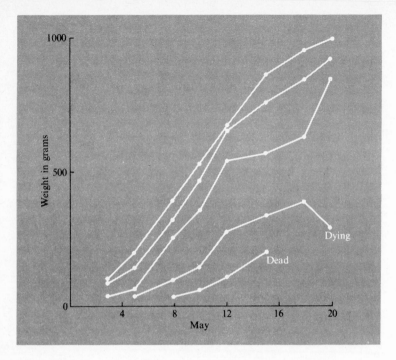

*Fig. 14:* Changes in weight in a brood of five nestling grey herons.

nestlings in a brood of five from hatching until they are almost fully
grown. Two nestlings hatched on 2 May, a third on 3 May, a fourth on
5 May, and the last on 8 May 1955. The sequence of hatching suggests
that incubation began gradually at about the time the third egg was
laid and that there was perhaps an interval of three rather than two
days between the laying of the fourth and fifth eggs. The first two
nestlings to hatch grew rapidly, as did the third until 12 May when its
rate of growth began to slow down. The fourth bird grew more slowly
and died of starvation on about 20 May. Almost immediately after this
the third nestling put on weight and soon caught up with the first two,
indicating that the presence of the fourth nestling restricted the growth
of the third. The last-hatched nestling lived only a week and never
stood a chance, as can be seen in Fig. 14.

The events depicted in Fig. 14 show what happens in a heron's nest
when there is a large family to feed in a year like 1955 which was poor

for fishing in the Thames Valley. In such years broods of only two or three nestlings are usually completely successful but in broods of four or five one or two young die of starvation. Occasionally large broods are actually less successful than small ones because the stresses imposed by a large family affect all of its members. But in good years when the weather is dry and fishing better the larger broods are more successful, and many parents may raise four or even five young. We can now see the advantage of staggered hatching. If all the nestlings hatched together and were the same size competition in the nest might result in the loss of the entire family. When hatching is staggered, however, there are always small nestlings in the family. Because of their size these are poor competitors when the parents return with food, and it is only when food is plentiful that they gain weight satisfactorily and survive. The first-born nestlings nearly always survive.

Recoveries of ringed birds in July and August show that the probability of survival depends a little on the size of the family when the bird leaves the nest: birds from larger families have rather less chance of survival than those from smaller families, presumably because in moderate-to-good years large families all survive, and the individuals tend to be in slightly less good shape, and perhaps slightly lighter in weight, than those from small families. The difference is not great but it is enough to show that large families, although apparently sometimes successful in the nest, may not do so well later on.

The parallel between what happens in a heron's nest and what happens to human families with insufficient means to support a large number of children is at once apparent: deprivation and lack of success in later life are as much a feature of large human families as they are of large families of herons.

We are now in a position to summarize and to ask questions. Herons are more or less solitary except when they come together in the early spring to breed at traditional sites. The size of the colony depends on the carrying capacity of the local environment which can essentially be defined in terms of the availability of fish and similar food. The importance of availability rather than abundance of food is striking and has to be viewed in the context of the heron's method of catching prey. Survival of nestlings depends on the parent's ability to obtain food, which in turn depends on the weather and perhaps other factors not at present known. In poor years, many nestlings die, most deaths occurring in large families. Once they have left the nest there are further heavy losses during the first year of life, but once a bird has lived a year its prospects of breeding several times are good. A population of herons in an area like the Thames Valley fluctuates from year to year, and there is a fall in numbers after a hard winter, but the fluctuations are nowhere near as great as potentially possible. It looks as if so far as herons are concerned the carrying capacity of the environment can be understood entirely in terms of the availability of food.

But why should an essentially solitary species become social and breed in colonies for a small part of the year? Would it not be better if single nests were scattered over the countryside so that available resources could be exploited and shared more efficiently? There must be some value in social breeding, not fully understood, but possibly connected with a scarcity of safe nest sites, and possibly because the birds require a certain amount of mutual stimulation early in the breeding season. Herons have elaborate courtship behaviour which takes place at the colony in early spring and time and time again you can see that courtship and mating in one pair will stimulate others to follow suit. There is also the question of how a heron breeding for the first time settles in an established colony. One-year-old birds visit the colonies, sometimes build small nests, but only on rare occasions do they attempt to breed. They seem to participate in the affairs of older birds, joining in with some of the courtship displays and often following adults to and from the feeding grounds. It is almost as if there is some form of apprenticeship during which by imitation and by trial and error the young birds acquire experience of colony life. Finally why is it that after generations a colony is suddenly deserted, and how are new colonies started? No one knows, and clearly there is still much to learn about the ecology of herons.

In Britain the grey heron has no serious competitors, unless we include anglers, and its distribution and abundance seem to be determined by the availability of suitable fishing grounds where it is not disturbed. The increasing use of lakes and rivers for leisure and the growing threat of water pollution are likely to have an adverse effect on its numbers. It is a handsome bird, familiar but not especially common, and its family life as at present understood can perhaps teach us something about the balance between resources and population.

## Periodical insects

Most insects have short life cycles lasting a year or less. A few species have life cycles lasting two or more years, but the individuals in a population are out of phase, so that adults appear every year. As a consequence the various environmental factors that may limit numbers, including food supply, predation, and parasitism, operate every year. Insects which depart from this annual pattern by having life cycles of two or more years with the entire population in phase are said to be periodical.

The best-known periodical insects are three species of cicadas of the genus *Magicicada* found in the eastern United States. The immature stages (nymphs) feed by sucking juices from the roots of deciduous trees and all but a few are at the same age at a given time during their long period of growth. Every seventeenth year (thirteenth in the south of the range of distribution) the nymphs emerge from the ground, moult into adult cicadas, lay eggs, and die, all within a few weeks. Each of the three species has a 17-year and a 13-year form. At a single locality where all three occur they are synchronized, although in different parts of their range they are out of phase. When the adults appear they are extremely abundant and conspicuous, much more so than related non-periodical species living in the same areas. Cicadas are large insects and the sudden flush in numbers attracts many predators, particularly birds, but they are so abundant that the predators become satiated and many cicadas survive. The beneficial effects on predator populations of increased numbers of prey are lost by the next emergence in thirteen or seventeen years time. This is the presumed explanation for the periodicity; it is a matter of flooding the market with food at long intervals and then quickly withdrawing the supply: no predator could become dependent on periodical cicadas.

Two species of cockchafer beetle, *Melolontha*, are periodical over most of their range in Europe, emerging as adults every three years (both species) in France and Switzerland and every four or five years (different species) further north. The oak eggar moth, *Lasiocampa quercus*, is periodical with a synchronized two-year cycle in northern Britain and Scandinavia. Whether or not the periodicity in these insects is a strategy that reduces the rate of predation is at present a matter for speculation.

## Naturally unstable populations

There is considerable evidence that populations are generally stable – tendencies for an increase in size are countered by higher death rates acting density-dependently. Several examples of density-dependence came to light in our discussion of herons. Population stability seems to be the rule but there are exceptions, among them plants and animals which at least temporarily undergo exponential increases in population size after being accidentally or deliberately introduced into new areas by man. There are also some naturally occurring exceptions, chiefly among animals at high latitudes and in dry areas in or near the tropics. In these areas there may be conspicuous fluctuations in numbers of a magnitude rarely seen elsewhere.

The Arctic and the Antarctic are characterized by a few abundant species of plants and animals. A somewhat similar situation exists in desert and semi-desert areas although here perhaps there are rather more rare species as well. If you go to Iceland in the summer you will

*Plate 7:* The south coast of Iceland. There are rather few species of plants and animals and the area is uncomplicated when compared with continental areas at lower latitudes. Almost the whole of Iceland has been severely grazed by sheep and the landscape has been much modified as a result.

be as impressed by the extreme abundance of a few species of sea-birds, moths, and midges as you will be by the lack of butterflies and the scarcity of all but a few species of song-birds. Likewise if you visit the dry savanna of Africa you will see vast numbers of a few species of weaver birds and white butterflies but many of the animals of the more humid tropics will be missing. It is in environments like these that certain populations become unstable and undergo big fluctuations in numbers.

Population instability manifests itself in two ways. There are populations which fluctuate regularly in size, a period of low numbers being followed at a regular and predictable interval by a substantial increase. This process continues year after year and fluctuations are so regular that they have been called population cycles. Although there is no disagreement that populations rise and fall, the regularity with which they do so has been questioned by some ecologists, and perhaps exaggerated by others, but all things considered it seems that at least for some species population cycles are a reality. Some of the species involved in cyclic fluctuations are associated with one another; thus a cycle in a population may lead to a corresponding cycle in its predators, a not unexpected event. Then there are populations which increase in numbers at irregular intervals that cannot be predicted with any degree of certainty. When this occurs there are often large-scale movements, called irruptions, of mobile animals like birds and winged insects. Irruptions become known as plagues if the organism in question is damaging to human health or to crops.

The best known cycles occur in lemmings (which are small rodents) in the Arctic tundra of Eurasia and North America. Lemmings attain high population densities every fourth year. When this occurs they sometimes undertake long-distance movements presumably because the local environment can no longer support them. Reports of lemmings throwing themselves off cliffs into the sea are probably exaggerated. Peak numbers are followed by a population crash after which lemmings may be so scarce that they are hard to find, but numbers begin to build up again and the next peak is reached about four years later. The lemming cycle is closely followed by a cycle in numbers of skuas, owls, and other predatory birds that feed on them. The predators do not really fluctuate markedly in numbers but rather move into and concentrate in areas where the lemmings are common and disperse when they become scarce again. There is no evidence to suggest that the predators feed on the lemmings in sufficient quantity to be responsible for the crash in the population. More probably the crash is brought about by a shortage of food (lemmings feed on vegetation) and occurs when the environment has been damaged to such an extent that further increases in numbers are impossible, although it has to be admitted that it is by no means certain that substantial numbers of lemmings actually starve to death when the population is falling. But it does seem that when the tundra has been freed temporarily of the pressures exerted by the lemmings it begins to improve, which in turn allows a build up of lemming numbers, and so the process is continued. It has been suggested that the rise and fall in numbers is caused by an intrinsic physiological cycle among individual lemmings making up a population. But although there are undoubted changes in the physiology of several species of Arctic rodents which are correlated with fluctuations in numbers it is not clear whether these cause the cycles or whether the cycles affect the physiology. Lemmings are Arctic animals; would they still have cycles if they lived at lower latitudes? In other words is it the relatively uncomplicated environment of the Arctic which makes possible their cycles, or is it something about the lemmings themselves?

In Europe the sudden appearance in late summer and autumn of large numbers of unfamiliar birds has attracted attention for centuries. In Britain the crossbill and the waxwing are the species best known for their sudden appearance, but there are others, all of them breeding in continental areas to the east and north-east. The irruptions seem to result from an unusually good breeding season followed by a failure in a local food supply. The birds then fly south and west and in some years reach Britain. Irruptions are irregular and unpredictable and most of the species involved are seed-eaters. At high latitudes the abundance of the annual crop of seeds is notoriously variable from year to year and this variability is thought to be responsible. It is a matter of conjecture whether a significant number of the birds return to their place of origin but a few crossbills will remain and breed in areas they have reached after an irruption.

But the most spectacular and for man the most important irruptions are those involving locusts. A locust is any species of grasshopper whose numbers are liable to build up unexpectedly. Locust irruptions are associated with arid regions of the tropics and sub-tropics and do not often occur in temperate or in humid tropical regions. Most grasshoppers feed on living vegetation, especially grasses, and they are especially abundant in grassy savanna areas. Almost all grasshopper populations remain relatively stable like those of other organisms, but under certain conditions the several species known as locusts increase to plague proportions and undertake long-distance movements causing extensive damage to crops and other vegetation. In Africa there are three species of locust that are serious pests, and we shall discuss one of them, the desert locust, the most destructive insect in the world.

The desert locust normally occurs sporadically and in small numbers throughout much of the drier savanna and desert region of Africa south of the Sahara, in Arabia, and in Pakistan. At low population density it behaves like one of the numerous other species of grasshopper but periodically, often in association with unseasonal rainfall, and the subsequent growth of vegetation in arid places, its numbers rise rapidly. The Red Sea coast has long been known as a source of massive outbreaks of the desert locust, and for thousands of years since the beginnings of agriculture in Egypt the coming of the locusts has been feared. The desert locust is particularly destructive to cereal crops and with the spread of cultivation in the savanna region of Africa it is becoming more and more a potential source of disaster. Locust plagues affect the lives of millions of peasant cultivators as the invasion area extends through 110 degrees of longitude from West Africa to Assam, north to Turkey and south to southern Tanzania, an area of about 26 million square kilometres containing some of the poorest people in the world. This enormous area includes many independent nations occupied by people of a variety of political and religious persuasions, many of them hostile to one another, a situation that is bound to lead to frustration when attempts are made to control locust plagues on an international basis. But the essential feature of the area is that it is arid.

Changes in numbers of the desert locust are correlated with changes in the quality of the population. When locusts are at low density they exist in a state known as phase *solitaria*, but as they increase there is a transition to what is known as phase *gregaria*. It has long been realized that the key to understanding the build up of locust plagues lies in the nature of these phase transformations. Individuals from the two phases differ in several attributes, but especially in behaviour and coloration. Phase *gregaria* individuals tend to aggregate, they are more mobile, their behaviour is highly synchronized, and the development from egg to adult is quicker. The immature locusts (known as hoppers) are black and yellow while those of phase *solitaria* are green. Locusts in phase *gregaria* are therefore more active and social than those in phase *solitaria*, which amounts to the same as saying that they are more opportunist and have a better potential for swarming than phase *solitaria*. A population of locusts is converted from *solitaria* to *gregaria* by the stimulus of rising population density. When this occurs the vegetation may be depleted and if they are still in the hopper stage and unable to fly they march in large bands in a fixed direction until more food is located. Bands of hoppers join up and the whole process can quickly assume enormous proportions. Once they become adult they can fly and if food continues to be depleted they undertake long-distance downwind flights. Further breeding occurs and the process is repeated several times, each time on a larger scale than before; within a few years there may be hundreds of millions of individuals. At this point swarms of locusts begin to arrive in areas where they have been absent for years in such numbers that the local vegetaion, crops as well, may be virtually wiped out.

*Plate 8:* A swarm of desert locusts in Ethiopia. These insects may destroy the crops of millions of people in the drier areas of Africa.

A plague of locusts may last for years and then quite suddenly decline, the insects reverting to phase *solitaria*, and assuming their normal role as ordinary grasshoppers. Locusts are successful opportunists and a plague may gather so much momentum that it will destroy everything green over vast areas, yet its recession is just as remarkable and unexpected as its initial increase.

We should certainly not be lulled into a feeling of satisfaction just because we think we understand how the plagues build up, and we should not be too confident that it will always be possible to control plagues by spraying with insecticides. The desert locust has been responsible for untold human misery in the past and we may well not have seen the last of it. In June 1978 the Desert Locust Control Organization for Eastern Africa announced that in Ethiopia locusts were out of control. War in the Horn of Africa prevented spraying of breeding grounds earlier in the year. Ethiopians are already going hungry as a consequence of the war, and full-scale famine seems inevitable, especially if there is another drought. The Organization felt that unless concerted and expensive measures were taken locusts would spread to Kenya, Tanzania, and the rest of East Africa, and would ruin the crops of millions of people.

## Some rival theories of how populations are regulated

The origin and development of a locust plague can be viewed as an event initiated by the sudden appearance of suitable food in the form of green vegetation in an otherwise barren environment. The ability of locusts to transform from perfectly ordinary grasshoppers to highly mobile opportunists can be understood, but the causes of the recession of a plague are far from clear. It is unlikely to be associated with the availability of food, as it should always be possible for the locusts to move to new and unexploited places. We are thus left with an unresolved problem.

The balanced populations of herons and the events which occur in their nests can be interpreted in terms of density-dependent factors operating in response to a limited food supply. This is not to suggest that no other factors are involved but in herons it certainly seems that the key factor is the carrying capacity of the environment, defined in terms of the availability of fish in shallow water.

We have in this chapter stressed the availability of food as being the most important aspect of the environment regulating the size of animal populations. The requirements of green plants must be considered separately. Plants need a space to grow and to capture the radiant energy of the sun. They also require water and inorganic materials, but they are not mobile and cannot seek out sources of water and nutrients by moving. Hence competition in a population of plants is much more likely to be for a suitable space to grow. There is no reason why the density-dependent mode of population regulation believed to occur widely in animals should not also occur in plants. Gardeners will be familiar with weeds competing for space with flowers and vegetables and the removal of weeds or thinning of seedlings is such common practice in gardening and farming that it tends to be taken for granted. Plants suffer from weather more than animals because most animals are capable of seeking shelter. Severe and unexpected drought or frost results in the loss of many plants quite independently of their density, but on the other hand some spaces where plants grow are better than others, and if there is competition for these we would expect density-dependent factors to operate.

Many animals are preyed upon by other animals. Losses from predation occur density-dependently and in some situations the effects of predators may be more important than food shortage in controlling population size. In humid tropical forests where there is an abundance of living vegetation all the year round and where it is unusual to see signs of defoliation it is possible that animals living on this vegetation are not limited by food. Such animals are rarely common, and they are preyed upon by a variety of predators, leading some ecologists to suppose that plant-feeding animals as a group are not food-limited. We shall examine this question in more detail later and for the moment simply note that predators could potentially regulate populations in a density-dependent manner below the limits set by the availability of food.

There is another theory which accepts that population regulation is density-dependent and that the availability of food is of paramount importance, but proposes that animals (or at least some animals) regulate their own density below the carrying capacity of the environment. The theory is based on an interpretation of the significance of territorial and social behaviour and argues that the function of such behaviour is to enable members of a population through contact with each other to assess their own density and to adjust their reproductive rate accordingly. On this theory herons would through social interaction at the colony produce fewer young at high and more young at low density. The theory has never been properly tested and evidence for it is flimsy and can equally well be interpreted in terms of the direct operation of density-dependent factors. It might also be added that the theory does not seem to hold for human populations: the situation in India would seem to provide the necessary arena for deliberately limiting population growth, especially when we remind ourselves that man is the most social and at the same time the most territorial of all animals.

There are also ecologists who reject density-dependent factors and propose instead that most deaths occur independently of density. Climate is invoked as the main cause of fluctuations in numbers and, it is suggested, populations are regulated below the level set by the availability of food resources. Deaths from food shortage and predators, although of course acknowledged to occur, are not thought to be important in regulating overall numbers. Most of the evidence claimed to support the theory of density – independent regulation has been derived from work on insects, and there is no doubt that they suffer heavy losses attributable to climate. But whether climate is important in the long run, and whether it is more important than the strictly biological factors causing density-dependent deaths, is a matter of conjecture. The last word has certainly not been said and it would be unwise to be dogmatic in accepting or rejecting any or all of these rival theories.

# 4 Communities, food webs, and organic diversity

If an angler used to fishing in the Thames Valley were asked to make a list of the fish and other animals he encounters during his fishing trips his list would be almost the same as the one of food items taken by herons given in Table 4 (pp. 78–9). He would acknowledge that the roach is the commonest of coarse fish and that the pike is a common predator of other fish. He would frequently see water voles and moorhens, perhaps an occasional stoat, and from time to time dragonflies would perch on the end of his rod. He would be aware that vegetation in and around the water affects his fishing, and that heavy rain or unusually dry weather determine the kind of fish he catches; indeed he would sometimes blame the weather if he catches no fish at all. An angler's successes and failures are similar to the heron's.

A gardener knows that some plants grow well together and that some require special kinds of soil. Certain plants need more water than others, some require the addition of fertilizer, and many seeds will not germinate if planted too early or too late in the season. Sunshine is necessary to ripen tomatoes satisfactorily, rain is good for beans, and if there are long spells of sunny weather there is an increase in the abundance of blackfly and greenfly on the roses. The appearance of a garden reflects the personal likes and dislikes of the gardener; it also depends on the kind of soil and the climate of the area. Gardeners, like anglers, are optimists; they frequently try to grow more than the garden can reasonably be expected to sustain. An experienced gardener knows the limits of his own garden, and will adjust his activities accordingly, but a beginner will try almost anything and usually fail with most attempts.

Both gardeners and anglers are aware of seasonal change. Conditions for fish and flowers vary with the season, and angling and gardening have to be adjusted to these changes. Throughout the world seasonal changes in temperature, rainfall, sunshine, and day-length dominate the lives of all organisms, including man, although those of us living in cities may be aware of these changes merely as a talking point or in relation to a forthcoming weekend in the countryside.

The assemblage of plants and animals found in a river, in the garden, or in any other place is called a community, a term not to be confused with the community concept of sociologists which refers to human social structures and interactions. In the biological sense communities consist of a variety of species, many of them never seen by the average person because of their habits and small size. The combination of a community and its environment is called an ecosystem. Ecosystems are discussed in Chapter 5.

One of the main features of communities is the extraordinary intricacy of interactions between individuals of the same and of different species; we are here back to the question of who eats whom and who breeds with whom mentioned on the first page of this book. Virtually every species of plant is eaten by several species of animals, and some like oak trees support hundreds of different species, most of them insects. Flowers are visited for their nectar by bees, hoverflies, butterflies, and moths, and this ensures that pollen is transferred from plant to plant and from place to place. Without insects flowering plants as we know them could not exist and most of the familiar insects are dependent on flowering plants. Green plants are of central importance in all communities for they are the producers of food upon which all other organisms depend, and this is true whether we are considering a garden, a river, or the sea. Plants are also the most obvious organisms in a community, followed by the animals feeding directly on them, and then there is a huge variety of less conspicuous predators and parasites feeding on the plant-feeders and on each other. There are also decomposers of dead vegetation and dead animals, including bacteria, some insects, and earthworms. The combined efforts of these organisms eventually account for everything produced by the green plants, and there is no substantial accumulation of plant material.

### The species concept

Here we must digress a little and explain what is meant by a species, a term we have already used several times, but one that must be understood more fully now that we are discussing the inter-relationships of species in communities. For most purposes a species of plant or animal can be said to consist of similar individuals which actually or potentially are able to breed among themselves. Two organisms belong to different species if reproduction between them is physically, physiologically, or genetically inhibited. Hybrids between species sometimes occur, especially among plants and animals domesticated by man, but wild hybrids are exceptional. All species are given a scientific name made up of two parts, the genus and the species. For example there are several kinds of white butterflies which because they are structurally similar to one another are placed in the genus *Pieris*, but they do not breed together and are therefore designated as separate species: *Pieris rapae, Pieris brassicae*, and *Pieris napi* are the three species, commonly called cabbage whites, that occur in Britain. Other species of white butterflies structurally different from *Pieris* are placed in different genera.

In practice however the species concept is more complicated than this. If two similar but not identical groups of organisms live in separate geographical areas the possibility of breeding together does not arise because there is no physical contact between the two groups. They may therefore be of the same or of different species – there is no way of knowing for certain – and all that can be done is to make a guess based on their degree of similarity. Such a guess is not as arbitrary as might first appear, for it will be made by a person who knows the group of organisms well, but nevertheless it is still a guess. There are also organisms which do not reproduce by a sexual process and with these the question of breeding together does not arise. Asexually reproducing organisms are separated into species on the basis of structure in much the same way as organisms which occur in separate geographical areas.

Most of the organisms mentioned in this book are readily classifiable into species simply because we have deliberately chosen familiar plants and animals whenever taking an example, and we have not been too concerned with micro-organisms which present the biggest difficulties. Most of us have no trouble in deciding whether birds seen in the garden belong to the same or different species, but we may with justification hesitate when we look at brambles growing wild in the hedgerow. Brambles seem to occur in a bewildering variety and it is by no means certain which kinds breed together and which do not. But all we need acknowledge is that for some organisms the species concept is clear-cut while for others it presents difficulties, which can perhaps better be appreciated when we have discussed how species originate.

The differences between species result from evolutionary changes which are understood in general terms but difficult to verify experimentally. This is partly because the evolution of a species takes a long time, perhaps thousands of breeding generations, which means thousands or tens of thousands of years for the larger and more conspicuous plants and animals. In animals one possible mode of species-formation has gained almost universal acceptance among biologists. The process envisages a population becoming split into two or more groups which remain isolated for a long time because of geographical barriers. No reproduction between the separate parts of the population is possible and each can become adapted to the environment in which it is living. Very often there results a considerable accumulation of differences between isolated groups, sometimes to such an extent that after many generations the various parts of the original population are quite dissimilar in appearance.

If after a long time the geographical barriers disappear and the animals meet again several possibilities arise. One is that the animals have changed so much in isolation that they are no longer capable of breeding together. Since species are defined in terms of reproductive isolation we now have two or more species where previously there was only one. But it is possible that even a long period of separation may not result in sufficient differentation for interbreeding to be impossible when the populations meet again, and if this is the case there has been no species-formation. It is believed that the majority of species of animals and many plants evolved in this way, but the process is less convincing in some plants, which seem to have evolved by a special kind of hybridization in which the number of chromosomes is increased.

Chromosomes are minute thread-like structures incorporating the hereditary material, or genes, located in the cells of organisms; usually a species has a fixed number in each of its cells. Sex cells, whether sperm, pollen, or eggs, contain half the chromosome complement of the body cells of an organism, and at fertilization the normal chromosome number for the species is restored. In the special process called polyploidy, there is a doubling of the chromosome number following hybridization, and so individuals are produced with twice the number of chromosomes of either of their parents. The process can be rather more complicated than this, but the fact remains that there are numerous plants and some animals where the chromosome number is an exact multiple of their nearest relatives, and such species cannot normally breed with those with the lower number of chromosomes. The precise details of how polyploidy is achieved need not detain us but we might note that many cultivated plants have been developed by natural or artificially induced polyploidy.

There are other theories which claim to account for the origin of species but it is widely agreed that most animal and many plant species have evolved as a consequence of geographical isolation and subsequent meeting, and that speciation in plants in particular frequently results from polyploidy.

We shall be discussing the number of species present in communities and their various ecological roles and in doing so we must bear in mind that species are not fixed entities but are constantly being subjected to environmental pressures and are therefore liable to evolutionary change. Many of the pressures stem from interactions with similar species and result from competition for similar foods. Birds in a garden are clearly separable into their respective species, brambles in a hedgerow are more difficult, and there are many intermediate cases, all of which indicate that we must apply the species concept with caution.

One of the curiosities of nature that has intrigued people for centuries is why some groups of plants and animals contain numerous species and others very few. There is a story that J. B. S. Haldane once found himself in the company of some theologians who began teasing him about his lack of belief in God. One of them is said to have asked Haldane what he could infer of the nature of the Creator from the works of creation. Haldane's reply was, 'An inordinate fondness for beetles.' There are more species of insects than of any other group of organisms and more species of beetles than of any other group of insects. It is by no means clear why this should be so. About eighty per cent of the species of land animals are insects and there are more species of weevils (just one of the many groups of beetles) than there are species of vertebrates, including fish. This provides some idea of the enormous diversity of insects. It is likely that species of insects as yet undiscovered outnumber those that have been thus far described and named by biologists. If you were to collect samples of such insects as tiny parasitic wasps in the tropics you would find that almost all the kinds you collected were previously unknown to science. This is because there are so many of them and few people have collected tiny parasitic wasps in the tropics. But in groups like birds there are probably few undescribed species because birds are conspicuous and attractive and correspondingly well known.

Among plants the familiar flowering kinds make up the bulk of the species. Flowering plants and insects are mutually dependent, and one group would not be possible without the other. Their evolution has proceeded in a complementary way over millions of years to such an extent that an analysis of a terrestrial community is dominated by considerations of plant-insect relationships. Both insects and flowering plants are effectively absent from the sea; marine communities are thus fundamentally different from those on land and are based on algae, small floating plants, and a variety of invertebrate groups like crustaceans, worms, corals, and molluscs.

## Inter-relationships between species in communities

Green plants occupy a central position in all communities whether natural or man-made. Their abundance and diversity depends upon the quality of the soil or water, the climate, and the amount and seasonal distribution of light. A community supporting a large quantity of vegetation supports an abundance of animals and a community rich in species of plants tends to support a rich diversity of species of animals. The climate of the earth is continually changing and this affects the vegetation occurring in a particular area. Millions of years ago the area which is now London was covered with tropical forest of the sort at present found in south-east Asia; this forest has long since disappeared and has been replaced at various times by deciduous and coniferous woodland, swamps, and ice fields. This kind of replacement is on a grand scale and takes thousands or millions of years, and we only know it has occurred because of the presence of fossil leaves, seeds, and pollen which can be identified and whose age can be accurately determined.

Each species plays its part in maintaining the essential structure of the community and every individual needs a space in which to live and reproduce, and access to resources. An individual animal requires food, shelter, and a place to hide from enemies; a plant requires light, water, and nutrients. The way in which an individual exploits the environment to satisfy these needs, or in other words, its role in the community, is often called its niche. The term niche implies not only a place to live but, especially in mobile animals like vertebrates and insects, the manner in which an individual operates successfully in familiar surroundings. Success in attaining the necessities of life depends on the individual's ability to respond to environmental stimuli. Thus an individual butterfly starts the day by seeking suitable flowers from which it can obtain nectar. It recognizes the appropriate flowers by their colour and scent, avoiding those that look or smell wrong and those that are utilized by another individual. A butterfly is quick to recognize potential enemies and takes appropriate evasive action. Possible mates are recognized by their appearance and smell and an individual will produce and disperse its own smell in order to attract a mate. A female is expert at recognizing the correct plant on which to lay its eggs and will examine plants repeatedly for the right stimuli, laying eggs only when it is satisfied that it has made the correct choice.

We sometimes speak of a person's niche to mean his role in life. When we do this we are not thinking especially of his ecological role but rather how he fits into his job and neighbourhood and the extent to which he 'gets along' with others (i.e. his place in the 'community' of the sociologist rather than the ecologist). In assessing a person's niche we are describing his abilities in terms of his surroundings or environment, but we do not normally go so far as to define a person's niche in terms of his capacities for survival and reproduction; we have substituted other criteria, but our assessment has many similarities to that an ecologist might make of an individual plant or animal in a community.

In Chapter 3 we discussed competition between individuals of a population, and explained how through competition density-dependent regulation of numbers is achieved. There is in addition the possibility of competition between species, and the more similar two species are the more intense the competition. In the early 1930s a Russian ecologist, G. F. Gause, observed that whenever there are several similar species in a community there are ecological differences between them. No two species seem to have precisely the same requirements. Essentially the same observation was made in the middle of the nineteenth century by Charles Darwin, but Gause was perhaps the first specifically to formulate the proposition that no two species found together are ecologically identical and that the differences between them result from and are maintained by competition. He was impressed by the observations of another Russian, A. N.

Formosov, on the differences in feeding behaviour between similar species of terns breeding on an island in the Black Sea, and in 1934 wrote:

The nests of the terns are situated close to one another and the colony presents a whole system. However, as regards the procuring of food, there is a sharp difference between them, for every species pursues a definite kind of animal in perfectly definite conditions. Thus the sandwich-tern flies out into the open sea to hunt certain species of fish. The blackbeak-tern feeds exclusively on land, and can be met in the steppe at a great distance from the sea-shore, where it destroys locusts and lizards. The common-tern and the little-tern catch fish not far from the shore, sighting them while flying and then falling upon the water and plunging to a small depth. The light little-tern seizes the fish in shallow swampy places, whereas the common-tern hunts somewhat further from the shore. In this manner these four similar species of tern living side by side on a single small island differ sharply in all their modes of feeding and procuring food.

The differences in the ways in which food is exploited by these terns are brought about and maintained by inter-species competition. By sharing the available resources it is possible for all four species to live in the same area. Gause verified his proposition by experiments with micro-organisms and many have since confirmed his results and conclusions: each species has its own ecological requirements which differ at least in part from those of related species.

Three species of leaf-eating monkey and four species of fruit-eating monkey live in the forests of southern Ghana. Of the three leaf-eaters, the red colobus feeds in the canopy, the black colobus in the middle and lower zones, and the olive colobus near the ground. Of the fruit-eaters, the diana monkey feeds in the canopy, the mona in the middle and lower zones, the spot-nose monkey near the ground (where it takes leaves as well as fruit), and the white-crowned mangabey mainly on the ground, specializing in fallen fruit, some of which may be accidentally dropped by monkeys feeding in the trees. Thus, as with the Black Sea terns, competition is minimized because each species of monkey takes a different range of foods in different places.

Interactions between dissimilar species range from such generalized feeding relationships as herbivores with plants, predators with prey, and decomposers with carcasses, to special associations sometimes involving total dependence of one species on another. An association between two different sorts of organism where one gains and the other is unaffected is known as commensalism. Such an association usually involves one participant attaching itself to the other and thereby reaching new food sources. Thus mites may be found on mobile bumblebees and beetles, *Hydractinia* (a plant-like animal related to sea anemones) grows on shells inhabited by hermit crabs, and remoras (fish with a strong sucker formed from part of the dorsal fin) attach themselves to sharks. *Hydractinia* obtains transport to muddy areas in which it could find no substrate for attachment and feeds on food fragments dropped by the crab. Remoras are carried to where sharks are feeding, release themselves to eat fragments from the sharks' meal, then re-attach themselves to the same or another shark.

A more familiar situation is parasitism in which one partner in an association benefits and the other is harmed. Parasites, unlike predators, are usually relatively small and rarely kill their host: they live at the expense of the host often by feeding on it over a protracted period of time. Parasites vary in the intimacy and permanence of their association with the host. Blood-sucking insects like mosquitoes and tsetse flies take a meal and then leave; fleas, ticks and lice attach themselves to mammals and birds and suck blood, but they are mobile and can if necessary move to another individual. These blood-suckers live externally to the host, causing irritation but little serious harm unless they are very numerous and remove large amounts of blood; but their indirect effect is enormous, since while feeding they often transmit disease-causing micro-organisms. Some female parasites attach externally but then develop into egg-producing bags within the host; among them are *Stylops* (an insect) on bees, and various tiny Crustacea on fish.

The most intimate associations are achieved by internal parasites which often lack structures concerned with feeding and movement. Adult tapeworms living in mammals' intestines consist of little more than an anchor and metres of egg-producing segments. Internal parasites usually have complicated life cycles, the young stages looking quite different from the adults, and often inhabit one or more intermediate hosts, one of which may act as a vector for transmission of the parasite to a new host. Thus the single-celled malarial parasite passes part of its life cycle in mosquitoes and is injected into man and other vertebrates when the mosquitoes suck blood; the parasite causing sleeping sickness in man is transmitted in a similar way by blood-sucking tsetse flies. The parasite that causes schistosomiasis, a debilitating and widespread tropical disease, passes its larval life in fresh-water snails from which a free-swimming form is released. This enters the skin of anyone wading or washing in infected water. The success of a parasite depends on the survival of the host; infestations heavy enough to cause the host's death are inefficient from the parasite's viewpoint.

In symbiotic (mutualistic) associations both participants benefit. The most familiar example is the relationship between flowering plants and their insect pollinators: both groups are completely dependent on each other. Lichens – ranging from the grey or yellow encrustations on walls to the feathery growth of 'Spanish moss' on trees in tropical forests – are an association of fungus and alga so close that they cannot be separated. The fungus obtains oxygen and carbohydrates from the alga, while the alga gains water and mineral salts from the fungus, as well as protection from desiccation and a means of attachment to the substrate. Ants of the genus *Pseudomyrmex* are symbiotic with various species of swollen-thorn acacia in Central America. The ants live and raise their broods of larvae inside the enlarged stipular thorns of the acacias and feed from specialized nectaries on the leaves. The tips of the leaflets are modified into unique structures, called Beltian bodies, which are rich in proteins and fats. These are harvested by the ants and are cut up and fed to their larvae. Swollen-thorn acacias produce new leaves throughout the year, even in areas with a pronounced dry season, so the ants are not only housed but are fed by the tree. The relationship is mutually beneficial because the aggressive worker ants keep the tree free of herbivorous insects, bite and cut vegetation touching the acacia, and clear plants from the area round the tree. Untenanted acacias are quickly defoliated or choked by vegetation and rarely, if ever, produce a seed crop.

Whether we consider predator and prey, parasite and host, or any other sort of interaction, we can view a community as an assemblage of species of plants and animals the diversity and abundance of which depends on the physical features and climate of the area in which the community is established. Each individual and each species is dependent to some extent on resources provided by other species, and unless disrupted by human activities a community can persist in a state of equilibrium year after year.

## Succession

If a cultivated field is abandoned, the original vegetation of the area
will not start to reappear immediately. There will at first be a vigorous
growth of weeds, followed by grasses and woody shrubs, and it may be
many years before the original community is re-established. As various
species of plants succeed one another, there are corresponding changes
in the species of animals present. Certain plants cannot become
established until the environment has been sufficiently modified by
other plants: moss normally found on bark cannot of course flourish
until there are trees. The sequential changes which take place are
called succession by ecologists, and the final association of plants and
animals that becomes established is called the climax. There are,
however, areas which, because they have been cultivated, are altered
more or less permanently so that the original vegetation never
reappears. Instead a new kind of community is developed quite unlike
anything that existed before.

The rate of succession depends on a variety of factors, the most
important being climate and the availability of suitable plants nearby
which can move in and colonize the area. You can demonstrate
succession on a small scale by placing some pond water in an aquarium
and suspending small squares of plain glass in the water. If you
examine the squares under a microscope every few days you can see
that algae are quick to colonize and that as time passes there is a
change in the species-composition of the algae. There will also be
changes in the species-compostion of small animals associated with the
algae, and after a time a spectrum of plant and animal life will become
established and will remain more or less unchanged for an indefinite
period.

The events occurring on glass squares suspended in pond water are
similar to what happens on a larger scale when a river is dammed and
an artificial lake is created. The terrestrial vegetation left in the flooded
area will begin to decompose, providing a source of nutrients for
starting the first stages of succession. As time passes and succession
proceeds, the water will develop a distinctive flora and fauna. The
early stages are often characterized by the presence of a few species in
great abundance, but then the diversity of species tends to increase and
a climax community usually possesses a greater variety of plant and
animal life than the successional stages. Succession in man-made lakes
and the possible climax communities have attracted a good deal of

attention from applied ecologists. Some tropical lakes become choked with aquatic plants soon after they have been created, making the water resources less beneficial to the human population than had been hoped. Indeed it is a sort of ecological game to predict the stages of succession and the resultant climax community. Fisheries biologists are interested to see what species of commercially important fish can flourish in a man-made lake, and this depends not only on the rate of succession but also on what species of organisms happen to become involved. By continually assessing the kind of succession taking place it is possible to say which species of fish should be introduced and when the introductions should be made: a mistake can sometimes have disastrous effects, even a total loss of fish stocks. In man-made lakes and in all other environments where succession is occurring, the species composition at first changes rapidly, then as time passes, more gradually, and when the climax is approached succession may be so slow as to be almost beyond detection.

The largest of Africa's man-made lakes is the Volta Lake in Ghana with a surface area of 8,500 km$^2$. It was formed by damming the Volta River about 100 km from its mouth, producing an irregularly-shaped lake filling the valleys of a much-branched river system. The dam was closed in May 1964 and subsequent events, as river became lake, have been monitored by the University of Ghana and the Ghana Academy of Science. For economic reasons there was little felling or burning of vegetation before the land was flooded. Soon after the dam was closed the abundance of dissolved nutrients from submerged vegetation caused a burst of growth of phytoplankton and a sudden increase of oxygen in the surface water. This was of short duration and organic decomposition in the deeper water reduced the level of oxygen to a point where large numbers of fish died.

Initially there was considerable growth of floating water cabbage, *Pistia stratiotes*, but in contrast to some man-made lakes, floating vegetation has never been a real problem. *Pistia* persists only in the most sheltered parts of the lake and even there is gradually replaced by sedges and other aquatic plants. Unfortunately, plants like *Pistia* provide food and shelter for the fresh-water snails that act as intermediate hosts for the parasite, which, as we have seen, causes schistosomiasis, and the disease has increased among people living near the shore.

The lake receives nutrients from the inflowing rivers to supplement those produced in the mud by decomposers. Availability of nutrients and of oxygen is enhanced by regular periods of water overturn and stirring which occur annually in the northern half and (for climatic reasons) twice-yearly in the southern half of the lake. Consequently the lake is favourable for fish, although maintenance of high productivity depends upon a continued and adequate influx of nutrients from outside. Once the lake had recovered from the initial destruction of fish caused by deoxygenation and decomposition, most of the original river species flourished for the first year, but then the character of the fish fauna changed. Elephant snout fish (mormyrids), typical of rivers, disappeared except from the mouths of inflowing rivers, and the large complex of characid fish was reduced to two species, perhaps because they normally breed in flowing water. But three species of commercially-important fish (*Tilapia*), scarce in rivers, are now abundant, presumably because of the increase in submerged vegetation and algae on which they feed. A mayfly, *Povilla adusta*, which as a larva builds tubes for shelter in rotting wood, has thrived; it feeds on algae and is itself a major fish food, forming an important link between microscopic plants and large predatory fish like the Nile perch. In the Volta Lake the economic necessity of leaving trees *in situ* on flooded land has benefitted resulting fish populations. The lake has presumably not yet settled down to a climax community, but already many lessons have been learned for application when tropical rivers are dammed in the future.

Heathland periodically demonstrates a series of successional changes in colonization. After burning, as happened so dramatically on the Dorset heaths during the 1976 drought, all above-ground vegetation may be destroyed, leaving only the fire-blackened, twisted trunks of the larger gorse bushes. By next season the first signs of new growth appear; bracken penetrates into the burnt areas by underground stems, heather shoots sprout from surviving roots, and wind-borne rose-bay willowherb and ragwort seeds arrive and germinate. Soon there are green bracken fronds everywhere, and around each new willowherb plant radiating underground rhizomes give rise to a cluster of daughter plants. The vegetation may take up to twenty-five years to revert to its original form, but the land does not remain bare for long. In many ways heathland cannot be regarded as a true climax community: to persist as such it must be pushed back into repeated succession either by burning or grazing; without this a patch of heathland would become woodland.

Indeed climax terrestrial communities are now something of a rarity, but it might be that man himself is in the process of creating climax communities. The countryside of England, with its pastures and arable fields, its hedgerows, and tiny patches of undisturbed woodland has persisted for generations and is effectively a climax community, although, it must be admitted, it is a community that nature would never have produced.

## Tropical rain forest and temperate woodland communities compared

Rain forest is best developed in South and Central America, West Africa and the Congo Basin, and in south-east Asia, while temperate woodland (which for our purposes will be oak woodland) occurs in western Europe, eastern North America, and parts of China, to name the more important areas in the northern hemisphere. Because both

*Plate 9:* The English countryside with its fields, hedgerows, and tiny patches of woodland, is possibly a man-made climax community.

types of community occur around the world there are striking geographical differences in species-composition from place to place: thus in South America virtually all the species of plants and animals are different from those that occur in African rain forest. Oak woodland in western Europe is similar to that in the eastern United States, except that there are fewer species of most groups of plants and virtually all groups of animals. For our purposes it is the structure of the communities that matters and not the occurrence of this or that species of plant or animal. We shall try and create a general impression, nothing more, and in doing so we shall have to make some rather sweeping generalizations.

It is simpler to have in mind particular areas than to generalize about communities that occur in many parts of the world and so we shall in the following pages keep in mind rain forest as it manifests itself in West Africa and temperate woodland as it occurs in southern England. The broad similarities and differences apply to other areas, but many points of detail are at variance, and it would be impossible even to begin to consider these. And before proceeding further we must admit that despite what is shown on world vegetation maps the two types of community under discussion hardly exist in their natural state in either West Africa or southern England. Both have been exploited, altered, and destroyed by human endeavour and there are now few places where it is still possible to see communities as they must have been before the advent of agricultural and industrial man. There are in England small patches of mature oak woodland planted about 250 years ago and believed to show many of the features of a climax community; likewise in West Africa there are areas of rain forest exploited for timber and allowed to regenerate which are probably similar to undisturbed forest, but it is essential to keep in mind the disrupting effects of man.

For much of the year the rain forest is wet and humid. Rainfall is high and conspicuously seasonal; in the wettest months of the year the forest is gloomy and hardly any sunshine penetrates to ground level. There may be 100 cm of rain in a month, more than twice the amount an English oakwood would receive in a year. The sequence of leafing and flowering and the life histories of animals are determined by seasonal rainfall. In some West African forest areas there is little or no rain in the dry season, which may last several months, but the humidity is high throughout the year, especially at ground level where because of the dense vegetation there is only slight air movement. The humidity is much lower in the canopy, especially on windy days in the dry season. Temperatures are relatively constant at 25–30°C, which is not hot, and there is a bigger difference between night and day than between seasons. If you walk in the rain forest you will not feel particularly hot, indeed it may seem cool in the shade, but the high relative humidity, often near 100 per cent, is striking and oppressive. At low latitudes incremental changes in day-length are small and have little effect on plant and animal life, in marked contrast to the situation at high latitudes.

The most impressive feature of the rain forest is the bewildering variety of plants and animals. In an oakwood, trees of the same species commonly occur side by side and only an elementary knowledge of botany is required to identify most of the flowering plants. In rain forest, however, it is unusual for there to be two individuals of the same species of tree in sight at the same time and many species of trees are relatively rare. Much experience is required before trees can be confidently named to species: those that look alike are often unrelated and those that look different from each other may be in the same genus and therefore closely related. Much the same is true of the animals, especially the insects, and superficial impressions as to what group a specimen belongs to are often wrong. A naturalist walking in an oakwood can usually identify with confidence most of the obvious plants and animals he encounters; there are few who can do the same in rain forest. There may be more than two hundred species of trees in a small patch of rain forest as compared with about ten in an English oakwood. A. R. Wallace, best known for association with Darwin in the formulation of the theory of evolution by natural selection, was one of the first to draw attention to the rich diversity of life in tropical rain forest, and in 1878 he wrote:

If the traveller notices a particular species and wishes to find more like it, he may often turn his eyes in vain in every direction. Trees of varied forms, dimensions and colours are around him, but he rarely sees any one of them repeated. Time after time he goes towards a tree which looks like the one he seeks, but a closer examination proves it to be distinct. He may at length, perhaps, meet with a second specimen half a mile off, or may fail altogether, till on another occasion he stumbles on one by accident.

An insect collector has much the same experience. He may be impressed by the variety of species but he will certainly not be impressed by their individual abundance.

Mature rain forest trees are tall but for their height their girth is not large. Trees of 50 metres in height are common, usually with a straight smooth trunk and no branches until near the top. It is as if the trees are for ever pushing upwards and once they reach the canopy they spread out and form a dense mass of foliage, which in many ways is a community in itself. Climbing plants and epiphytes are abundant: some of the larger older trees are covered with vines, orchids, and ferns. Strangler fig trees entwine themselves around other trees to such an extent that the trees themselves become inconspicuous and may often be killed. When a big tree falls a space is created and this is quickly colonized by a succession of plants until the tree's place is taken by another individual, usually of a different species. Leaf fall occurs in all months of the year, but fallen leaves quickly decompose, and there is little accumulation of humus in the soil; indeed the ground is often bare and stony with little vegetation. Many of the smaller shrubs below the canopy are saplings belonging to the same species as the big trees, although they seldom mature. Some trees have conspicuous buttresses like huge planks of wood, others are supported by stilts, and virtually all are shallow-rooted, although the roots spread over a wide area.

Animal life in the canopy is quite different from that on or near the ground. Tall towers have been built to investigate the ecology of the canopy but for obvious reasons more is known about what happens near ground level than high in the tree tops. Animal life is in every way more stratified in rain forest than in other land communities and is perhaps best paralleled by life in the sea or in lakes.

You can easily form a superficial impression that seasonal events in the rain forest are negligible, even though some times of the year are quite clearly wetter than others. But this impression is wrong; virtually every species of tree or plant has a seasonal cycle of leafing, flowering, and fruiting which presumably is triggered by environmental events, the nature of which (it must be admitted) is often unclear. And because the plants are seasonal, animals that depend on them are also seasonal. Many species of insects may occur as adults in all months of the year but for each there is a sharp peak of abundance associated with seasonal changes in the plants upon which they are dependent. The leaves of rain forest trees are tough and leathery and are largely unpalatable to plant-feeding insects except when they are new and soft. Many of the older leaves contain toxic chemical compounds

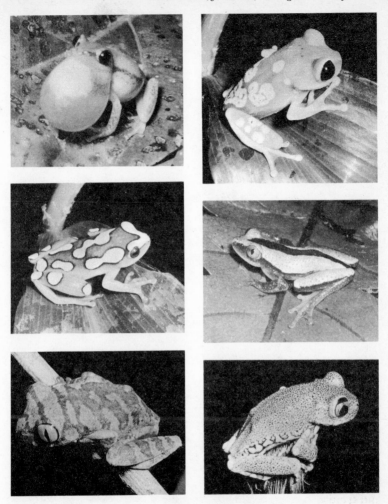

*Plate 10:* Species diversity in African tree frogs. Six of the many species are shown. Most conceal themselves and are hard to find but in the wet season forest regions are dominated by the sound of their calls. The top left picture shows a male calling with its throat inflated.

which discourage animals from eating them. It is certainly possible to form an impression of an abundance of food in the form of greenery, but the impression is wrong because so much of the foliage is inedible or even poisonous.

The rain forest is in terms of the diversity of species the richest and most complex type of community on land and is rivalled in the sea only by coral reefs along tropical shores. Ecological inter-relationships in the forest are complicated: thus an insectivorous bird may spend the day feeding on hundreds of species of insects, hardly ever taking two of the same species in succession, while its temperate counterpart would be faced with relatively few species, each of them more common than any of the rain-forest species. There is much scope for opportunism in the rain forest, especially among insect and fruit feeders. In West African forest the chimpanzee is one of the most successful opportunists. Chimpanzees move about the forest in search of food, travelling usually on the ground, sometimes in the canopy, and as the season changes they move from species to species of tree seeking ripe

*Plate 11:* The royal antelope, the smallest species of antelope in the world, smaller than a rabbit. It lives in West African forests, where it feeds on flowering plants growing on the forest floor, but it is rarely seen.

fruit. If there is a sudden emergence of winged termites, they may temporarily abandon fruit and become insectivorous; some individuals even scavenge from human refuse dumped near villages, and they frequently leave the forest to raid crops in adjacent farmland. Their diet is immensely varied, as it is with all animals which depend on fruit and insects. But animals that feed on the leaves of plants tend to be restricted to a few species that may be relatively rare, which may account for the fact that although there are many species of leaf-eating insects none is especially common.

Besides its complexity the rain forest also gives the impression of stability. It is not receptive to colonization by plants and animals from other communities and few of the species carried around the world by man have been established in rain forest. But if the forest community is disturbed by tree felling or cultivation, outside invaders soon gain a hold. Thus virgin forest contains few grasses but as soon as a pathway is cut grass spreads along it and becomes common; indeed if you try and study the forest community by staying on paths you will obtain a misleading impression of the flora and fauna. At the same time, forest plants and animals tend to survive badly outside their own community. In this respect nothing is more striking than the failure of the Congo forest pygmies, whose way of life is intimately adapted to their environment, to assume a new role as other people interested in cultivation alter and destroy the forest community. Pygmies are simply unable to come to terms with modern developments; their numbers are now much reduced; no one cares about them, and their future seems gloomy. Indeed their position is like that of some of the species of forest trees which cannot grow in isolation because they require the entire complex community in order to survive. Some of these trees are potentially of commercial importance, but attempts to grow them as pure stands in the absence of other species meet with limited success. Monocultures of trees are familiar in temperate lands, but we cannot expect to see vast areas of the tropics supporting trees of the same species. Silviculturalists often try to reduce tree species diversity in favour of desirable monocultures, and in tropical Queensland (for example) such attempts have sometimes been unsuccessful. In at least one species of timber tree germination is inhibited when the trees are grown together but not when they are grown mixed with other species. This is an example of density-dependence associated with a density-induced growth inhibitor, the chemical nature of which is obscure.

Even potentially mobile animals of the rain forest like birds and winged insects are sedentary compared with those of temperate woodland. Rain forest birds are not migratory; they tend to move around in flocks, called 'bird armies' by ornithologists, and rarely leave the forest community, although they are quite obviously capable of doing so. Insectivorous birds breeding in temperate woodland are largely migratory and some fly south to spend the northern winter in Africa. With a few exceptions they do not enter the rain forest community; of the many European birds wintering in West Africa only the wood warbler commonly enters the forest, the other species spend the winter in more open areas and cultivated land. It is as if the rain forest is already full and can provide no possibilities for the migrants.

Ecologists familiar with temperate woodland find themselves confused when they first explore rain forest. Initially they are unable to find much evidence of animal life and it is easy for them to form the impression of scarcity, which, however, is misleading. Unlike the animals of temperate woodland, rain forest species tend to be more restricted in their ecological requirements, they have (if you like) a narrower niche in the community; they are less common and more dispersed, and with the extraordinary variety of vegetation are better able to conceal themselves. Many species live in the canopy and hardly ever descend and will therefore not be seen unless a special effort is made.

Ants are an exception – they occur everywhere. Many of them are predators of other insects and we must suppose that coming to terms with the ants is a major problem for virtually all other insects and invertebrates. There are not only large numbers of ants but also many species; some nest high in trees and almost every leaf, flower, stem, and twig is covered with ants. Plant-feeding insects like bugs produce sugary secretions attractive to ants and in this way avoid being eaten. Indeed some plant bugs and butterfly caterpillars are attended and guarded by ants and are looked after in much the same way as people look after cows for their milk. Trees containing large ant colonies seem to suffer less from leaf-eating insects because the ants keep the leaves clear, and farmers in the forest region growing crops like oranges should not discourage ant nests as their presence results in less damage from the numerous insects that attack orange trees. Some plants, including some of the forest climbers, are equipped with special glands in the leaves that produce a substance attractive to ants. The ants are thus encouraged to frequent the leaves and at the same time they remove eggs and caterpillars which might otherwise inflict damage and inhibit growth of the plant.

Almost all the remaining patches of undisturbed rain forest are surrounded by cultivated land and there is a gradation from the assemblage of plants and animals created by man into the true forest. The border between undisturbed and disturbed forest is by no means clear although an experienced observer can readily pick out indicator plants that will tell whether man has altered the natural community. Where forest edge occurs naturally there is an abrupt transition: the forest suddenly stops and is replaced by woodland mixed with grassland. An even more impressive transition occurs when the forest reaches a lake, a river, or the sea. The edge is of considerable interest because it possesses many of the properties of the inaccessible canopy and provides a situation where the canopy flora and fauna may be examined without the need to ascend to the tree tops. Areas of transition are called ecotones by ecologists; they occur in all communities and their interest is that they tend to contain mixtures of species from two or more communities, and they are as a result rich in species-composition. It would of course be equally valid to consider ecotones as separate communities; indeed we should not define the limits of communities too precisely.

And now let us turn to oak woodland. The few remaining patches in southern England are maintained for timber, for shooting interests, or for conservation. England lost most of its natural woodland centuries ago when oak was in demand for ship-building. Luckily there was a good deal of re-planting and we can thus gain some idea of what the original community must have been like. Compared with rain forest, oak woodland is dominated by one or two species of trees and as many as half of the larger trees usually belong to a single species; the comparable figure for rain forest is about five per cent.

Oak woodland does not have a distinct canopy fauna although a few species, like the purple hairstreak butterfly, are restricted to the tree tops. But unlike the rain forest there is usually a distinct undergrowth including plants like bracken, brambles, bluebells, and grasses. The trees are not as tall as those of rain forest but for their height their girth is greater. The bark is rough, the roots go deep into the soil, buttresses and stilts are absent, and epiphytes and climbers unusual. When a tree dies and falls over or when a large branch drops off it remains on the

*Plate 12:* A termite mound standing about 150 cm high and containing millions of termites organized into an intricate social system. Termites are primary decomposers of dead vegetation. Termite mounds provide an environment for the establishment of several species of plants and trees and they contribute significantly to vegetation associations in parts of Africa.

ground for years. It is tempting to ascribe the slow rate of decomposition to low temperatures, but more probably it is the absence of wood-eating termites that is responsible. In the tropics termites are the main decomposers of wood, and it is risky to construct wood buildings and fences and to leave timber lying around. Is it possible that the British navy which for centuries relied upon oak for ship-building would not have been nearly so important in world history if England had the termite fauna typical of tropical countries?

Temperate woodland usually has a deep litter of dead leaves and other decomposing vegetation and the soil is rich in humus, in marked contrast to rain forest. The leaves drop from the trees in autumn and growth during the winter is minimal. Flowering and fruiting occur at predictable times of the year, and temperature and incremental changes in day-length provide the environmental triggers for seasonal events. Rainfall is not important and oakwoods flourish in areas where the rainfall is so slight that if it were in the tropics it would result in a dry savanna community.

Oak trees support a rich variety of animal life. The eggs of several species of moths are laid in the autumn on the twigs and when the buds burst next spring the eggs hatch and the caterpillars feed on the new leaves. In some years caterpillars defoliate the oaks to such an extent that the trees are forced to produce a new crop of leaves later in the season. Defoliation is partly a question of timing: if the eggs hatch late and the leaves are already partly grown damage to leaves is less conspicuous, but if hatching occurs early, before the leaves are well developed, defoliation is more likely as the caterpillars are ahead of leaf growth. This is why in some years defoliation is obvious but in others it is scarcely visible. Later in the year as the oak leaves mature they become tough and produce toxic compounds which make them unpalatable to caterpillars, and by July there are rather few caterpillars feeding on the leaves. The sudden flush of caterpillars in the spring leads to exploitation by a variety of predators, especially birds, which use them to feed their young. Defoliation of the kind frequently seen in English oakwoods is unusual in rain forest.

Many of the animals of temperate woodland are dormant in the winter. Insects hibernate either as eggs, larvae, pupae, or adults, and reappear in the spring; many insectivorous birds are migratory and spend the winter in the tropics, reappearing and breeding in the spring. The animals of oakwoods may be characterized by their ability to exploit sudden seasonal changes in the availability of different foods: resources appear and disappear with conspicuous regularity, and the most successful species are those that are adapted to these seasonal changes.

As already mentioned species of oakwood animals are commoner than those of rain-forest animals. This suggests that inter-relationships in an oakwood are simpler, and indeed this may be so, but nevertheless an oakwood community is immensely complex, and we are only just beginning to understand how it functions.

## Grassland

World vegetation maps show extensive areas of prairie grassland in the interior of the North American and Eurasian continents and of savanna grassland with varying densities of trees in southern South America, central and southern Africa, and Australia. The present extent of habitats dominated by grass is rather different from that shown on world maps, since human activities modify grasslands just as surely as they do forests: there is undoubtedly more grassland in the world today than there was two thousand years ago, and this is entirely the result of the widespread destruction of woodland and of grazing the land by domestic animals. Compared to most plants, grasses are extremely palatable to herbivorous animals. Without grazers there would probably be no grass: we can therefore speak of grassland either as an example of 'permanent succession' or as a climax community maintained by grazing. The importance of grass is enormous: as a species man is largely dependent for food on the seeds of about half a dozen species, especially wheat, rice, and maize; without cereal crops the human population would necessarily be much smaller than it is now.

Nowadays large areas of temperate grassland are given over to cereal cultivation; tropical savanna grasslands are rather less modified by twentieth-century man, but this situation is changing. Rainfall determines the extent to which the savanna is wooded and the type and growth-form of the trees. All savanna areas are subject to a dry season of varying severity during which there are extensive fires, some caused by lightning when the dead stems become tinder-dry, but more commonly started by pastoralists to clear dead vegetation and promote the growth of new grass shoots palatable to livestock. Some of the characteristic plants of savanna are those able to withstand fire, which suggests that burning is ancient and occurred even before man discovered the use of fire about half-a-million years ago. Grass fires reach high temperatures but move quickly, and the heat penetrates little more than two cm below the soil surface. Fire-resistant savanna trees have thick bark, small leaves, protective bud-scales, and deep roots. The grasses are adapted to aridity, the leaves become tough and unpalatable with age, and they put out underground stems which enhance their spread and tide them over the dry season and the burning. With the first rains there is a sudden flush of new and tender grass which is exploited by antelope and other grazing animals which produce young to coincide with the increase in the availability of palatable food.

It is probable that savanna is contemporaneous with man. There is historical evidence that much of Africa that is now savanna or desert was once well-wooded if not actually forest. Shifting cultivation involving the burning of trees and the clearing of vegetation as practiced all through the tropics results in the conversion of forest to savanna. Clearance for cultivation on a significant scale started about 8000 years ago in Africa and major changes in vegetation have occurred only in the last 3000 years. As the forest retreated and the soils became impoverished by intensive cultivation, pastoralists moved in and introduced the custom of annual burning to improve grazing for their cattle, thereby creating conditions in which antelope, elephant, and other big mammals could increase in numbers and range. Were it not for the presence of tsetse flies which transmit the disease trypanosomiasis, cattle could have become more common in Africa than they are today: there would as a consequence be more savanna, less forest, and even worse problems of trampling, erosion, and desertification.

Once savanna grasslands are established by the interaction of climate, cultivation, grazing, and burning, other grass-dependent species move in and flourish. These include grasshoppers which feed on the leaves and stems of grasses, and seed-eating birds such as weavers. There are also changes in the soil and its fauna resulting in a delicately balanced community maintained by grazing and burning. At one place in Mali grazing has been restricted and the savanna has become woodland again. Buttressed trees of the kind typical of forests occur scattered through the relatively wet savanna of the Ivory Coast and extend into the dry areas of Upper Volta and Mali, providing further evidence that much of the African savanna has at least in part been created by man.

How many of us realize that the wild-looking grass-covered mountains in the west and north of Britain are a recent development? These areas were once forested, but the trees were cleared and sheep introduced which have since maintained a grassland climax community. Much the same has happened in New Zealand, Iceland, and parts of Australia and North America. The North American prairie, which in its heyday supported over fifty million bison, was a climax community maintained by grazing, possibly aided by fires started by man. Thus while grasslands are characteristic of particular regimes of temperature and rainfall, their development is inextricably interwoven with the evolution of grazing animals and the agricultural history of man. The most curious ecological feature of grass is its palatability: this is the key to understanding the nature of grassland communities throughout the world.

## Food chains and food webs

If we tried to draw a diagram of the possible feeding relationships occurring in rain forest or temperate woodland we would need an enormous piece of paper and our diagram would be so full of question marks and uncertainties as to be almost meaningless. All plants are eaten by animals of one sort or another and these animals are in turn eaten by others, and so on. There may be two hundred species of trees, a hundred species of birds, and perhaps ten thousand species of insects in a small area of rain forest and the feeding relations between these as well as between all the other plants and animals are beyond reasonable prospects of description: there are millions of possibilities, and only relatively few are known.

If we take as a temperate example the caterpillars of the winter moth (one of the common species feeding on oak leaves) we could draw a possible feeding relationship in the form of a food chain, as follows:

oak leaves → winter moth caterpillars → great tit → sparrowhawk

The arrows show the direction along which food is being transferred: the sparrowhawk kills great tits which eat winter moth caterpillars feeding on oak leaves. The trouble with the diagram is that it oversimplifies and does not show, for instance, that winter moth caterpillars are eaten by other species of birds as well as by wasps, spiders, and numerous other invertebrates, and that sparrowhawks kill other species of song-birds which may or may not feed on winter moth caterpillars. Food chains are therefore much more complicated than our diagram would suggest. How then can we represent what is happening?

We can begin with a generalization that is valid for all communities and starts with the assumptions that food chains commence with plants (whether living or dead) and that plants are consumed by herbivorous organisms, usually but not always animals, and that these animals are eaten by others. We can then write:

$$\text{green plant} \rightarrow \text{herbivore} \rightarrow \text{predator}$$

which is a suitable simplification. In the example involving the winter moth caterpillars there are two predators in the chain, and indeed this is common, although there are rarely more than four.

Food chains as we have drawn them are linear but in reality feeding interactions in a community are not linear and are more intricate than can be shown by drawing a chain. For this reason it is more satisfactory to draw what is called a food web, although even this is a simplification.

Fig. 15 shows a possible food web for a pike living in the River Thames. We are assuming that the pike is a big one, weighing perhaps 4 kg. There are many possible ways of drawing the web involving different groups of organisms; all of those shown in Fig. 15 are eaten by a variety of other species. At the base of the food web there are aquatic plants which by photosynthesis convert radiant energy from the sun into plant tissue. The plants are eaten by small animals, including crustaceans, tadpoles, and insects, and these in turn are eaten by bigger animals like roach, large aquatic beetles, and frogs. We have not in Fig. 15 allowed for the possibility of frogs eating their own tadpoles, which no doubt occasionally occurs. The roach, beetles, and frogs provide food for predatory fish like perch, and all (including perch) are eaten by the pike.*

We have specified that the pike is a big one, and it is interesting to examine a possible food web for a small pike, just a few centimetres long. This pike, if it is lucky, will grow up to become a top predator but during its early life it is further down in the web, as shown in Fig. 16, where it is put in the place occupied by aquatic beetles in Fig. 15. Like

---

*As shown in Table 4 pike are eaten by herons but individuals weighing 4 kg would be too big for a heron to manage.

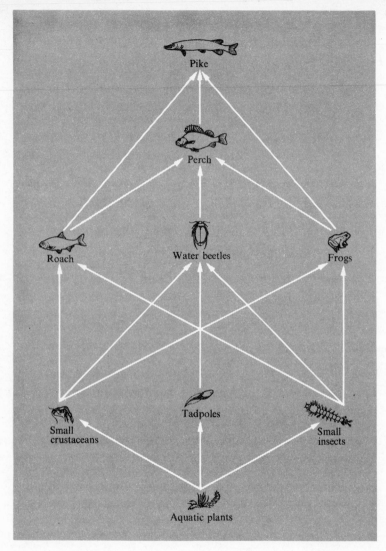

*Fig. 15:* Some possible feeding relations of a large pike.

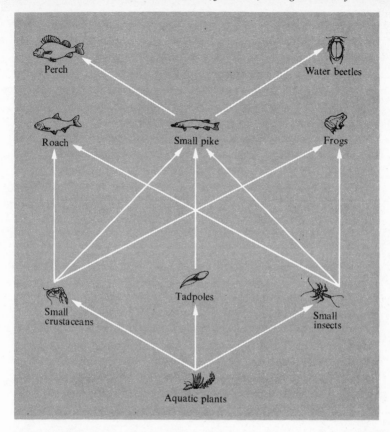

*Fig. 16:* Some possible feeding relations of a small pike.

the beetles it feeds on small items including crustaceans, tadpoles, and insects, and is itself eaten by perch and even by large beetles.

The point of showing two diagrams for the pike is to emphasize that we cannot be too rigid in trying to slot an organism into a food web. An organism's position will depend on several factors, not the least of which are its age and size. Thus insects may be herbivores when larvae and predators when adult; much depends on the life history of the species and on the circumstances. Even though some animals can without hesitation be classified as herbivores, a great many are at least occasionally predators.

Consider an adult butterfly in a rain forest community. It may take nectar from a flower, in which case it could be classified as a herbivore since it is feeding directly on a green plant, but within a second or two it may suck juices from a decomposing animal or drink from urine, which places it in a different category altogether. Generally speaking those animals that feed directly on green leaves are more conservative in their range of possible foods while those feeding on other animals, fruits, the nectar of flowers, seeds, and decomposing organic matter have a much wider range of possibilities. For these reasons it is impossible to draw a food web for a complete community and we must also bear in mind the large number of species present in even the simplest communities. As an instructive exercise try drawing a food web for your family based on your last meal.

## Organic diversity

The essential characteristic of a community is provided by the variety or diversity of species of plants and animals it contains. No two areas are exactly alike in species-composition but there are broad trends which can be defined and measured. One way to assess diversity is to make a list and to estimate the relative abundance of species that occur in the community. This is not entirely satisfactory because species are not all equally different from one another in either structure or ecological requirements but no one has come up with a better method, although there are various mathematical formulations, beyond the scope of this book, by which diversity can be computed with a fair degree of precision. A more serious problem encountered in diversity estimates is the complexity of communities which presents a huge and frustrating task to anyone bold enough to attempt a complete analysis of all species present. Consequently it is common for ecologists to confine their attention to selected groups of plants and animals which are easier to find and identify than others. In this way it is possible to compare species diversity in different parts of the world, in different communities, and at various times of the year.

Taking the world as a whole the most obvious trend, one which we have already mentioned in passing, is for species diversity to increase from the poles to the equator. In virtually all groups of organisms (some exceptions are known) there are more species in the tropics than in temperate or polar regions; indeed there is a gradient in species diversity correlated with latitude. There are other trends, for example temperate North America is richer in species than western Europe, and there are invariably fewer species on islands, especially remote islands, than on continental land masses, but these differences are nothing like as great as the latitudinal gradient.

About 136 species of butterflies occur in Michigan (latitude 42–48° N.) and about 720 in Liberia (latitude 5–8° N), two areas of roughly comparable size. About 20 species of butterflies may be recorded in an English suburban garden while a West African garden may produce nearly 300 species. Apart from the greater number of species many tropical butterflies are relatively rare while in temperate regions each species tends to be relatively common, although there are of course exceptions in both environments. A total of 219 species of ants has been found on a Ghana cocoa farm compared with 67 species on abandoned fields in Michigan. And so we could go on: these are just two examples, but curiously enough the reasons for increased species diversity towards the tropics have not been satisfactorily explained even though the phenomenon has been extremely well documented for over a hundred years. Animals depend on plants and strictly herbivorous species tend to be restricted to a narrow range of plant species. Since there are more species of plants in the tropics than in the temperate regions this alone could explain the greater diversity of animals, but it is unlikely to be the whole story, and in any case does not explain the high plant diversity in the tropics.

Tropical areas also tend to support a great variety of communities. Thus in tropical countries like Uganda there is almost every conceivable kind of terrestrial community from low-altitude rain forest to high-altitude tundra. In the tropics areas above 2000 metres support flourishing communities, whereas in the Arctic and much of the temperate region they are too cold. But the real problem is why even within a small area like a garden there are more species in the tropics than in the temperate regions.

If the number of individuals of each species in a sample of plants or animals is counted it will be found that there are a few common and many rare species. This distribution of species abundance is encountered again and again whether we are looking at flowers in a field, trees in a forest, or insects taken from a trap. A common species in one community may be rare in another apparently similar community and most rare species are common in at least some communities. Table 5 shows the frequency of species of *Acraea* butterflies caught, marked,

*Table 5* Relative abundance of *Acraea* butterflies in a garden near Freetown, West Africa, during 27 consecutive months

| Species | | Number of individuals | Relative frequency (per cent) | |
|---|---|---|---|---|
| Acraea | egina | 1589 | 23·56 | |
| | eponina | 1401 | 20·78 | 60·59 |
| | circeis | 1096 | 16·25 | |
| | pharsalus | 503 | 7·46 | |
| | natalica | 497 | 7·37 | |
| | bonasia | 461 | 6·83 | |
| | caecilia | 301 | 4·46 | |
| | lycoa | 274 | 4·06 | |
| | zetes | 230 | 3·41 | |
| | parrhasia | 165 | 2·45 | |
| | quirina | 149 | 2·21 | |
| | camaena | 27 | 0·40 | |
| | rogersi | 18 | 0·27 | |
| | encedon | 12 | 0·18 | |
| | jodutta | 9 | 0·13 | |
| | admatha | 6 | 0·09 | 3·37 |
| | pentapolis | 3 | 0·04 | |
| | alciope | 2 | 0·03 | |
| | perenna | 1 | 0·01 | |
| | terpsichore | 1 | 0·01 | |
| | Total | 6745 | | |

and released in a garden near Freetown in West Africa. This group of butterflies is common in tropical Africa where they are especially attracted to gardens which they enter in order to feed from flowers. *Acraea* butterflies are easily caught with a butterfly net, they are not difficult to identify, and they can quickly be marked on the wing with a spot of fast-drying ink. The sample was taken over twenty-seven consecutive months and all individuals, rare or common, were caught and recorded. The species obtained are listed in order of abundance. As shown there are twenty species but just over sixty per cent of the sample is made up of only three while the ten least common species comprise less than four per cent of the sample. A total of 6745 butterflies was caught, two of the species once only, and four others less than ten times each. In other parts of West Africa some of the rare garden species are common; thus in forest near the garden *Acraea quirina* is the commonest member of the group but in the garden itself it made up only 2·2 per cent of the total butterflies taken, and in nearby rice fields *Acraea encedon* is the commonest species yet in the garden was taken only twelve times.

*Acraea* butterflies feed as adults on the nectar of flowers, and although their overall abundance in the garden may be at least partly determined by the availability of flowers this is unlikely to be the cause for some species to be common and others rare. More probably numbers are determined by other factors, among them the availability of larval food-plants. The larvae of each species are confined to a narrow range of plant species, some of them locally common, some scarcer, but there is no obvious correlation between food-plant abundance and butterfly abundance, although it must be admitted that this would be difficult to ascertain as there is no easy way of assessing food-plant abundance. There is indeed something of a

*Plate 13:* A mated pair of *Acraea encedon*, one of the twenty species of *Acraea* butterflies found in a garden near Freetown, West Africa.

mystery as to why there is so much variation in the abundance of different species of *Acraea* butterflies in this garden but the pattern described occurs in almost every group of insects and in every community.

Leaving aside latitudinal gradients and other geographical trends, it seems that the greatest diversity of species occurs in the older well-established communities where fluctuations in population size are slight and where there is long-term stability. It would therefore seem that as more and more land is converted by man to monocultures there must result a fall in diversity in world communities which could lead to the instability of many populations and outbreaks of certain species which until now have been relatively rare.

## Weather and climate

We are all obsessed by the weather: casual meetings between acquaintances tend to start with a brief exchange about the rain, the sunshine, the fog, the cold, and nearly always we agree the weather is unusual for the time of year. Selective memories of hot summers or snowy winters years ago lead to a feeling that the weather is 'not what it used to be' and extreme events such as nuclear explosions or supersonic bangs are sometimes invoked as causes for want of any rational explanation. Improved forecasting techniques have generated increased demands for accurate predictions of short-term changes. Instantaneous world-wide communication of catastrophes stemming from drought, flood, and hurricane bring pictures of the accompanying misery into comfortable, insulated homes. We know more about what is happening around the world than we did a few years ago and this alone can give the erroneous impression that the world's climate is less stable than it used to be. If, however, we examine documented history for widely different parts of the world a sufficient understanding of weather patterns should emerge to allay our fears.

First we must draw a distinction between weather and climate. The climate of an area is determined by latitude, altitude, proximity to the sea, and prevailing winds; it is defined in terms of the average, and in some years, even several successive years, may depart from the average in rainfall, sunshine, or temperature. The climate helps to determine the vegetation and fauna of an area and accounts for seasonal patterns of growth and reproduction in plants and animals. Temperate deciduous trees shed their leaves in autumn and remain bare until the following spring, even though some days in winter are warmer and sunnier than some summer days. The climate of a particular place may not remain constant. In the geological past the global wind patterns that produce areas of high and low pressure and determine climate have moved both north and south; the polar ice caps have been both more and less extensive than now, alternately locking up and releasing vast quantities of water. Such episodes in geological history have produced long-term changes in climate.

Short-term fluctuations in climate occur more frequently than directional trends. In the seventeenth century Britain experienced sharp, dry winters and cool, moist summers, and some people believe that these conditions are returning. Even if this is the case, average temperatures would be only about 1·5°C below those experienced today. A small drop in average temperature would have some economic impact on the viability of upland farming and would lead to increased sales of warm clothes, but there is no reason to suppose that we are about to enter another ice age. Instrumental records date back only three hundred years and this is the explanation for the climatic extremes recorded in Britain recently: the coldest winter since 1740 (1962–3), the driest winter since 1743 (1963–4), the mildest winter since 1834 (1974–5), and the driest sixteen months on record (April 1975 to August 1976). Such extremes are often associated with a particular type of circulation in the atmosphere and consequently may persist for a time, but there is no reason to think of them as part of a long-term trend. Spells of variation such as these have led to the adage 'Britain has weather, not climate', which is the same as saying that fluctuations in weather are more apparent than climate. We are familiar with our own weather and our lives are adjusted to what we can reasonably expect. Rainfall that is normal and manageable for a farmer in Chad would constitute a disastrous drought in Britain; it depends on what you are used to: one man's desert is another man's Garden of Eden.

Recurrent drought is undoubtedly a major factor causing famine in the Sahel and must be considered in the perspective of the climatic history of the countries bordering the southern Sahara. About 15–20 thousand years ago there was a southwards shift of the climatic and vegetation zones of Africa and the Sahel became very dry; ten thousand years ago the situation was reversed and the area became wetter and warmer and remained so for a long time, despite a dry spell about six thousand years ago. Rapid desiccation set in about five thousand years ago and destroyed large areas of vegetation. During the past five centuries there have been climatic fluctuations in the Sahel but no general trend has emerged. Thus from the sixteenth to the eighteenth centuries it was wetter than 'normal', then followed a change towards the present dry conditions; but from 1680 onwards there were several severe droughts, and the present aridity was interrupted by a wetter period in the late nineteenth century. All of this means that the climate of the region is variable and unpredictable; no one can predict the likelihood of rain, and the fauna and flora of the region are adapted to withstand unreliable rainfall.

The recent history of Botswana also underlines the need to accept that weather is rarely predictable. The climate of Botswana is semi-arid to sub-humid with an annual rainfall of 25–65 cm falling between November and March. The land is predominantly grassy savanna and supports large numbers of cattle, sheep, and goats. Colonial reports on the occurrence of drought between 1890 and 1965 show that in 31 years the rains were late or absent and that in seven of these years there was severe drought. Crops were badly affected in 19 years and in three of these there were locust plagues which in Africa are often associated with unusual weather. In eight years cattle suffered badly. But prior to 1978 there were about five years of good rainfall and no immediate problem was evident: a visiting agricultural expert on a two-year contract to help in drought relief programmes could be forgiven for concluding that he had mistakenly been sent to a green and pleasant land.

A recent World Meteorological Organization document points out that a most important factor in understanding drought is that it is a supply and demand phenomenon. Drought depends on the density and distribution of plant, animal, and human populations, and how the land is used, not just on rainfall deficiency. There is a great deal of truth in this sentiment.

Evidence of a different sort implicates man as causing a slight but significant trend in the world's climate. It has been estimated that the destruction of vegetation that accompanied pioneering agriculture in the period 1860–90 released fifty per cent more carbon dioxide into the atmosphere than the total fossil fuels burnt up to 1950. Increased carbon dioxide in the atmosphere exerts a 'greenhouse effect' and traps heat radiated from the earth's surface. This would have produced a

0·5°C rise in the average world temperature in the late nineteenth and early twentieth centuries.

Natural communities show great resilience to abnormal weather. Over thousands of generations plants and animals have become adapted to fluctuations around the normal. Britain's 1976 drought had a dramatic impact on man, his crops, and his livestock; reservoirs dried up, field-grown vegetables withered, and by August cattle were being fed from winter stores. The response of the natural flora and fauna was noteworthy, but few cases of catastrophe were apparent. Some trees such as oak and birch shed their leaves prematurely, an obvious strategy for water conservation, and (curiously) horse chestnuts flowered in the autumn. Plant-feeding insects, aphids in particular, experienced a population explosion early in the summer, as did their hoverfly and ladybird predators. It was the best year for butterflies since 1947, proving that we had been wrong to assume that the apparent decline in butterfly numbers in recent years had been caused by over-use of pesticides and herbicides in the countryside. But by the next year things were more or less back to normal and two years later the drought was forgotten.

Thus in places as widely separated as the Sahel, Botswana, and Britain the weather is unpredictable. Perhaps these places are subject to particular variability because of their position relative to areas of atmospheric disturbance and associated wind patterns, but they are by no means exceptional. Even the seemingly stable equatorial rain forest is subject to variations around its climatic norm. In their day-to-day activities plants and animals are well-adjusted to changes in weather. But people throughout the world continue to complain and fail to understand that although we know a lot about climate our knowledge of weather is rudimentary.

## The seasons

We have already made passing reference to seasonal change in communities and we have now arrived at a good juncture to discuss the subject in more detail. Seasonal changes affect virtually all organisms in all communities and because of this cannot be neglected in an analysis of community structure.

There are two ways of looking at seasonal events. We can identify what are called proximate factors which are in many animals detected by sense organs in such a way as to provide information about the time of year. Birds in temperate woodland are stimulated to start breeding by the increasing day-length and rise in temperature in spring; those breeding in rain forest do so in response to changes (increases or decreases) in rainfall. Each species will respond somewhat differently from others and no two are identical in their reactions to outside stimuli. We can also identify ultimate factors which determine why an organism breeds at a particular time of the year. Ultimate factors unlike proximate factors are therefore concerned directly with the survival value or adaptive significance of seasonality. Proximate factors tell us how an organism knows the time of year and ultimate factors tell us why certain times of the year are better than others for breeding, migrating, hibernating, or indeed any other activity associated with the changing seasons. The distinction between these two sets of factors is not always entirely clear but if you think of birds in temperate woodland it is easy to envisage that they are responding proximately to day-length and temperature changes and ultimately to increased availability of food and nesting sites provided by the growth of vegetation in spring.

In temperate communities each species of plant or animal tends to reproduce only at a restricted season of the year. In the tropics a forest bird (or any other organism) is likely to breed at the beginning of the rainy season but unlike his temperate counterpart the tropical ornithologist would not be astonished to discover an occasional nest outside the normal breeding season.

Seasonal change tends to be taken for granted in temperate areas: everyone is aware of the progression of the seasons, and gardeners and anglers (who have special interests in the seasons) watch for changes and plan their activities accordingly. In fact you can buy many gardening books that will tell you what you should have done last month. There is a repeating pattern of change but there are also variations from year to year: we are all familiar with the occasional hard winter, late spring, or exceptionally fine summer, and indeed exceptions to expected weather are a pervasive topic of conversation. It is therefore not worth spending too much time discussing seasonal changes in temperate plants and animals. But in the tropics seasonality is less well understood and many people who have not been there are under the impression that the climate is the same all the year round and that most plants and animals go about their activities without regard to the time of year. We shall therefore give just a single example of how the seasons affect one group of tropical organisms and note that the pattern described occurs, with variations, in most other groups of plants and animals. We shall consider hawk-moths, a discrete group of large, mainly night-flying insects found in many parts of the world but especially common in the tropics.

Fig. 17 shows the average number of hawk-moths caught each night in a light trap operated for twelve consecutive months at Freetown in West Africa. The average number of moths each night remained more or less constant from October to March and then in April there was a sudden increase in numbers, and a peak was reached in May and June. The increase was caused not by the sudden appearance of species that

*Plate 14:* A convolvulus hawk-moth feeding from a spider lily flower. The moth's tongue is 15 cm long and curls up when it is not feeding. Pollen from the flower's anthers (the four vertically oriented structures) becomes attached to the wings of the hovering moth and is transferred to the stigma (the long horizontal tube in the photograph) of other flowers as the moth moves around and feeds. The structure of both moth and flower are mutually adapted and one would not be possible without the other. The picture was taken at night in a garden at Freetown at the beginning of the rainy season when hawk-moths are abundant and when the spider lily comes into flower.

until then had been absent but by an increase in abundance of species already flying. In July there was a rapid fall to the previous level. This pattern of seasonality is repeated year after year and the sudden increase (which is delayed until May in some years) is proximately stimulated by the occurrence of the first heavy rain following the long dry season. The rainfall figures for each month are shown in the lower

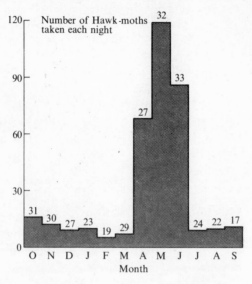

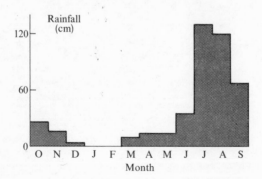

*Fig. 17:* Average number of hawk-moths taken each night in an insect trap compared with rainfall, at Freetown, West Africa, between October 1968 and September 1969. The numbers at the top of each column in the upper histogram are the numbers of species recorded each month.

histogram in Fig. 15. As can be seen it was dry immediately before the rise in numbers, but the fall occurred as the heavy rain developed in July and August. Thus hawk-moths are most common at the time of maximum change in rainfall after the dry season and not when rainfall is heaviest. The numbers given at the top of each column in the histogram are the number of species taken in each month. They remain relatively constant compared to the variation in abundance. In March there were on average only seven moths each night but during the month twenty-nine species were recorded; in April when there were sixty-eight a night only twenty-seven species were recorded. In other words species diversity expressed as the number of individuals in relation to the number of species was highest when the moths were least abundant and lowest when they were most abundant. This seasonal trend of species diversity occurs in several groups of tropical insects and is the reverse of what happens in temperate areas. Comparable figures obtained in England indicate that the number of individuals and the number of species rise and fall together; the peaks of individuals and of species are reached at the same time in mid-summer, and diversity and abundance are correlated.

Many plants flower at the beginning of the Freetown wet season and since adult hawk-moths feed on nectar it seems likely that this is the most important ultimate factor involved in the seasonality, but it is also possible that the production of new leaves (stimulated by rainfall) provides a suitable environment for egg-laying and for the survival of young caterpillars.

## Migration

Migration is the regular, usually annual, movement of a population to and from a breeding area. In some groups (birds, for example) the same individuals participate in both movements while in others (most insects) individuals move to the breeding area and their offspring move away. Migratory animals belong to mobile groups that are capable of long-distance travel. Birds and winged insects are the best-known migrants, but there are also migratory mammals, fish, and even toads. In a sense migration is part of the reproductive process and is affected by the same ultimate factor: the provision of optimal conditions for breeding. The exact nature of these conditions depends upon the kind of animal involved but nearly always food is paramount. Similarly, the proximate factors triggering migration depend on the kind of animal:

in many species a migratory movement is initiated by changes in daylength, temperature, or rainfall.

Swifts arrive in northern Europe from tropical Africa in May. Their sudden appearance is one of the signs of summer but they do not stay long and most have gone by early August. Swifts nest in colonies under roofs of buildings, often in ventilator shafts, and are nowadays thoroughly adapted to life in towns. They feed entirely on insects and tiny spiders caught in the air and are incapable of feeding on items on the ground or in vegetation. The female lays two or three eggs and the young are in the nest and being fed by their parents in July, the best month of the year for aerial insects. If the weather is fine and sunny and there is not much wind, the young swifts receive adequate food from their parents, but in spells of rough weather their growth is retarded and some die of starvation. Swifts arrive, breed, exploit the seasonal peak of aerial insects, and then disappear to more productive places in Africa. Why do they come here at all? The answer is that in July there are more insects in the air in northern Europe than in even the most favourable African environments.

A parent swift brings a compact ball of between 300 and 1000 insects and spiders each time it arrives at the nest to feed its young. The whole ball is passed to one nestling and there may be a struggle, a smaller nestling receiving food less often than a larger one. A brood of swifts is fed about 40 times a day and may receive about 20,000 insects and spiders, a phenomenal total as each food item must be individually seen and caught by one of the parents. Young swifts are in the nest for about 43 days, and, allowing for the food that the adults must take for themselves, a pair of swifts and its brood probably consume nearly a million insects and spiders during July.

The North American monarch butterfly is the best known migratory insect. Most of the eastern North American population winters in one small area of Mexico. Here millions cluster in the trees and roost in a semi-dormant state. The butterflies move north in spring and there are several breeding generations until autumn, when the return flight to Mexico commences. The most spectacular butterfly migrant to Britain is the painted lady which in some years is extremely abundant and in others rare. The centre of its distribution is the Mediterranean region of southern Europe and North Africa. The painted lady cannot survive the winter in Britain and all individuals seen in the summer are immigrants or the offspring of immigrants.

Migratory species are of particular interest in the study of communities because they are temporary members also occupying niches in other communities. It could be argued that the swift is a European bird that spends most of its time in Africa or an African bird that breeds in Europe. Migrants are characteristic of regions with a markedly seasonal climate and their strategy is to move in and exploit a seasonal flush of food. Most migratory birds are insectivorous and most migratory insects exploit successional plants (including weeds) which can become temporarily extremely abundant. Migration can thus be viewed as an alternative to winter dormancy: instead of hibernating mobile animals move elsewhere and remain active.

## Community structure

We have been concerned in this chapter with the various ways in which organisms are organized into communities, how they exploit the available resources, and how they respond to the environment. The restraints imposed by other members of a community and by the environment normally prevent any one species from becoming excessively common. Communities develop by passing through a series of successional stages and the resulting climax depends on the characteristics of the environment, especially on the climate. If a community is disrupted by human intervention, succession begins afresh and the new climax eventually established will probably be different from the original. Natural communities are relatively stable and it appears that the more species present the greater the likelihood of long-term stability.

We have been speaking mainly of the relationships between species and in the next chapter we extend these relationships to include the non-living part of the environment, paying particular attention to the acquisition and transfer of energy within communities and the ways in which nutrients are obtained and utilized.

# 5  Ecosystems and how they work

We must now develop a broader approach to ecology and introduce the concept of the ecosystem. A community together with the non-living part of the environment in which it is established and all the varied interactions therein is called an ecosystem, a term which is self-explanatory. The most important attribute of an ecosystem is its independence of external sources of matter and energy, other than light from the sun. Another is its capacity to circulate material, including water and other inorganic compounds and elements necessary for life to persist. An ecosystem can be looked at from several levels: it is justifiable to regard the whole of the living world as one huge ecosystem, but more commonly and more usefully we speak of the rain-forest ecosystem, the oakwood ecosystem, and so on, but no limits to size and complexity can be laid down, and it is perfectly correct to speak of an ecosystem centred around a single plant. The essential point in introducing the concept is that it includes everything that contributes to the maintenance of life within a specified space and time.

## Production in ecosystems

It has long been known that green plants are incapable of sustained growth in the absence of sunlight. On the other hand, plants can grow with nothing other than air, water, mineral salts in solution, and light. Unlike animals they do not require food in the sense normally understood, and yet plant material contains an abundance of energy, as you can see when you burn a piece of wood and produce heat and light. The chemical energy in plants is derived from the sun's radiant energy by a conversion process called photosynthesis. The process is remarkable in a number of respects, its most important ecological feature being that once the sun's radiant energy has been converted it remains stored in the form of chemical energy, being liberated only when used by the growing plant, or when the plant is consumed by another organism. Photosynthesis is unique to green plants, some red and brown algae, and a few bacteria, and so far no one has been able to replicate it in the laboratory.

Absorption of light energy requires the presence of pigments which are located in tiny structures in plant cells called chloroplasts. The most important pigments (there are others) are the chlorophylls which give plants their distinctive green colour. Clusters of chlorophyll molecules, called quantosomes, are arranged on membranes within the chloroplast. Under experimental conditions, photosynthesis can occur in illuminated suspensions of chloroplasts but not in extracts of chlorophyll. Light 'excites' chlorophyll molecules causing the emission of high energy particles which are used to build molecules rich in chemical energy, but there is more to it than this: photosynthesis is a highly complex series of reactions.

The first part of the photosynthetic process, called the light reaction, takes place in the quantosomes and involves the conversion of radiant to chemical energy and the splitting of water molecules. The oxygen formed by splitting water is given off by the plant and this is probably the original source of all oxygen present in the atmosphere. The hydrogen atoms from split water molecules are taken up by a hydrogen acceptor called nicotinamide adenine dinucleotide phosphate (NADP). The second part of photosynthesis, called the dark reaction, requires no light, but is dependent on the continuous supply of the products of the light reaction. In the dark reaction, which takes place in the watery fluid surrounding the membranes within chloroplasts, chemical energy formed in the light reaction, the hydrogen carried by NADP, and carbon dioxide give a molecule containing carbon, oxygen, and hydrogen. This is a rapid but complex cyclical process in which many intermediary substances are formed; it results in the production of a sugar which can then be converted into other sugars, starch, and other carbohydrates.

The discovery of the existence of separate light and dark reactions led to an understanding of how light intensity, carbon dioxide concentration, and temperature affect the rate of photosynthesis. It is the dark reaction, dependent upon an adequate carbon dioxide supply, that is speeded by an increase in temperature, while at low light intensities the rate of the light reaction limits the rate of photosynthesis as a whole. In summary the process of photosynthesis may be represented by:

$$\text{carbon dioxide} + \text{water} \xrightarrow{\text{light}} \text{carbohydrates} + \text{oxygen}$$

The productivity of an ecosystem is measured in terms of the rate at which chemical energy is produced during photosynthesis. The total amount formed in a given time is called the gross primary productivity and the amount left over after utilization for life processes by the plants is the net primary productivity; it is only this latter that is available as food for animals. The total amount of plant material present at any particular time is called the standing crop, or biomass, and is usually expressed in terms of dry weight per unit area; the dry weight is simply the weight of material after the water has been removed.

Animals like cows that feed directly on green plants are sometimes called secondary producers because they convert plant material into a form which can be used by other animals (especially man); but it is easier and equally correct to refer to secondary producers as primary consumers. Both terms are used by ecologists and they have the same meaning. The real producers or creators of organic material are the green plants, and all other organisms are dependent on them for food.

This may all seem rather technical but we are simply stating in rather formal terms something which is well known to farmers and gardeners. If a gardener plants a row of bean seeds they will eventually grow and produce beans which may be harvested and eaten. At the end of the growing season the gardener is left with his crop but there are also other parts of the plants – leaves, stems, and roots – left over, all produced by photosynthesis. Each individual plant has used some of its energy to keep itself going, but its net production is left over, and of this only the beans themselves are of interest to the gardener.

From what has been said it will be evident that ecosystems must contain a means of primary production, and therefore will almost always contain green plants. Without plants, ecosystems cannot exist as there is no source of organic materials, which is the same as saying there is no food. A rabbit infested with parasitic worms and covered with fleas is not in itself an ecosystem as the rabbit cannot sustain itself without feeding, but if we add grass we have an ecosystem centred around the rabbit. An aquarium containing water plants is an ecosystem, as the plants produce organic materials through photosynthesis. If you keep goldfish in an aquarium in which water plants are growing it is posssible for the aquarium to form a balanced ecosystem and to sustain itself indefinitely. It will probably require the addition of some small organisms to feed on the plants and to provide food for the goldfish; and it will also require organisms like bacteria to decompose the dead plants as they accumulate, because otherwise the aquarium will soon become choked with plant material. In practice it is not easy to establish a balanced aquarium; it can be done, although most people provide additional food for the goldfish and periodically clear out excess plant material. The ecosystem of the aquarium need not include fish: all that is basically needed are green plants and light, and of course decomposers. The plants themselves produce the necessary carbon dioxide when they break down stored carbon compounds and release energy during respiration.

Light, carbon dioxide, water, and salts in solution are all that is required for plant production to occur, but the rate of productivity varies with the availability of these necessities, except perhaps carbon dioxide, which is normally available in sufficient quantity. Available light varies with the season of the year, with latitude and with the weather, and in all areas of the world some seasons of the year are better for plant production than others. Most people know that in the northern winter plant production is at a minimum and many plants hardly grow at all. The rates of many biological processes are affected by temperature and plant production is no exception, although there is variation between species of plants in their ability to grow at different temperatures.

Water, too, is limiting, and especially in dry tropical areas where water quickly evaporates and is lost, lack of water may prevent growth in conditions that are otherwise suitable. Land plants require considerable amounts of water, not only for photosynthesis, but for transpiration which involves a large uptake and a large loss of water, and also because water is required by plant tissues for metabolism. No plants exist entirely in the air and all land plants require a substrate in which to support themselves and through which to acquire water and salts in solution. Large areas of land are unsuitable for the majority of plants because they are too dry or too cold. There are areas in the polar regions and on tops of high mountains where plant life is impossible because it is too cold, and there are deserts where it is too dry. Water is not of course a limiting factor for aquatic plants. Some like those we grow in the aquarium are attached to the bottom, but many float. Photosynthesis is not possible in deep water because light does not penetrate far, and even in the clearest ocean most photosynthesis is restricted to the layer of water near the surface. Small floating plants in the ocean and in fresh water are called plankton but the term is also used for any floating organisms, including animals. The plants concerned are minute, most of them single-celled and invisible to the naked eye; they occur in countless millions and at their most abundant impart a distinctive blue-green colour to the water near the surface.

Net primary productivity, net primary production, and biomass have been estimated for all the major world ecosystems, and the figures are summarized in Table 6. All the estimates are subject to wide variation and we must remember that many of the ecosystems listed are being drastically altered by man. Man probably consumes about 1/200 of the annual photosynthetic yield, a figure which is another approximation, but it is certain that man's consumption is increasing exponentially. The biggest average biomass occurs in tropical forests. This is because the trees are long-lived and grow large and there is

*Table 6* Estimated productivity, production, and biomass of the major world ecosytems (Adapted from R. H. Whittaker, *Communities and ecosystems*, Macmillan, 1970)

| | Area (million $km^2$) | Average net primary productivity (dry g $m^{-2}$ $year^{-1}$) | World net primary production ($10^9$ dry tons $year^{-1}$) | Average biomass (dry $kg$ $m^{-2}$) | World biomass ($10^9$ dry tons) |
|---|---|---|---|---|---|
| Lake and stream | 2 | 500 | 1·0 | 0·02 | 0·04 |
| Swamp and marsh | 2 | 2000 | 4·0 | 12 | 24 |
| Tropical forest | 20 | 2000 | 40·0 | 45 | 900 |
| Temperate forest | 18 | 1300 | 23·4 | 30 | 540 |
| Boreal forest | 12 | 800 | 9·6 | 20 | 240 |
| Woods and shrubland | 7 | 600 | 4·2 | 6 | 42 |
| Savanna | 15 | 700 | 10·5 | 4 | 60 |
| Temperate grassland | 9 | 500 | 4·5 | 1·5 | 14 |
| Tundra and alpine | 8 | 140 | 1·1 | 0·6 | 5 |
| Desert scrub | 18 | 70 | 1·3 | 0·7 | 13 |
| Desert and rock | 24 | 3 | 0·07 | 0·02 | 0·5 |
| Agricultural land | 14 | 650 | 9·1 | 1 | 14 |
| Total land and freshwater | 149 | 730 | 109 | 12·5 | 1852 |
| Open ocean | 332 | 125 | 41·5 | 0·003 | 1·0 |
| Continental shelf | 37 | 350 | 9·5 | 0·01 | 0·3 |
| Estuaries | 2 | 2000 | 4·0 | 1 | 2·0 |
| Total ocean | 371 | 155 | 55 | 0·009 | 3·3 |
| World total | 510 | 320 | 164 | 3·6 | 1855 |

much woody material. Ocean biomasses are much lower because planktonic plants do not require woody structure for support. The important difference between the forest and the sea is that although in the sea net primary productivities are fairly high, because of rapid consumption and decomposition and lack of woody structures, the biomass at any particular time tends to be low.

It is tempting to speculate that the stages of succession described in Chapter 4 would lead to increasing productivities and this is often but not always the case: a climax community may or may not have the maximum productivity for the conditions of soil, light, and weather for the area in which it is established. Climax should perhaps be characterized as a state in which there is equilibrium between production and consumption, and not necessarily in terms of productivity.

We now know roughly how much organic material the world can produce in a year and roughly how much is left available for utilization by other organisms, including man. We have some idea of the differences in yield between the main world ecosystems, and this knowledge, first acquired intuitively and by trial and error when cultivation began thousands of years ago, is the basis for our present agricultural policies and our hope for feeding the ever-growing population of the world. Modern agriculture is striving to find ways of increasing productivity, and of course the question arises as to how long improvement can be kept up. We should remind ourselves that the world is a closed system and we must make the best of what is available and try what is possible.

Energy entering an ecosystem by the process of photosynthesis is eventually lost in the form of heat; none is recycled. Organisms have to 'work' in order to live and this involves a loss of energy. Work done by active animals is easy to see and we are familiar with the fact that we lose heat when we work. Plants also work as they grow and build up a store of chemical energy, and there is a continual loss which has to be replaced. Fig. 18 is a diagram showing the relationships between the acquisition of the sun's energy by green plants and the loss of energy by plants and animals. The diagram emphasizes that plants produce the energy in a form that can be used by animals, and that without plants animals could not exist. Every time an animal eats plant material some of the energy is converted into animal tissue and some is lost either as heat or as indigestible waste products. The same is true when an

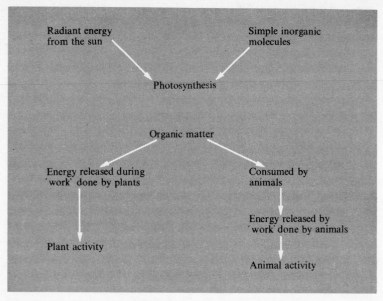

*Fig. 18:*  Energy inflow, photosynthesis, and energy loss in plants and animals.

animal eats another animal: there is continual wastage in the form of heat and as undigested matter.

Plants and animals are remarkably inefficient in their utilization of available energy. It is possible to measure the energy content of sunlight, and the energy content of plant material can be found by burning the material completely and measuring the heat given off. The ratio of the energy in plant material to that of incident sunlight gives a measure of the efficiency of plants, and this turns out to be low, not more than one per cent for most plants, but with much variation. Similar estimates can be obtained by measuring the energy content of animals and the plants they eat. As might be expected growing animals are more efficient at utilizing the energy of food than adults; an animal eating just enough food to keep alive and to keep its weight constant has an efficiency of near zero. This means that unless an animal is growing it consumes an amount of energy equal to what it needs for day-to-day maintenance. Farmers are interested in the efficiency with which cattle can convert grass into meat for human consumption. They are also interested in the optimum number of cattle that can be kept for a period on an area of pasture. If there are too many cattle the pasture will be ruined by over-grazing, and this creates problems for the farmer, but if there are too few the grass produced is under-exploited and the farmer loses money because he is not making the best of the available resources. Good farmers know how many cattle may be kept on a piece of pasture without destroying the pasture for future use.

## Trophic levels

Every time an organism consumes another organism there is a loss of energy. Armed with this knowledge we can predict that plants will be more abundant than animals, and that animals that feed on plants will be more abundant than animals that feed on other animals, and so on. No one would seriously disagree with this prediction – ordinary observation of our surroundings confirms its validity – but the concept of abundance must be viewed with caution. We are clearly not speaking of the number of species because there are more species of animals than there are of plants. We must therefore try and define abundance in other terms. There are various possibilities. First, there is the number of individuals, regardless of their species; but this creates problems as some individuals are large and some small, and in many plants and some animals the concept of the individual is ambiguous. This is especially true of plants that reproduce vegetatively. How for example would you decide how many individual grasses there are in a field? The task is virtually impossible and the answer would be nothing more than an unreliable guess. Next we can consider biomass as a possible measure of abundance. Biomass seems more promising as it takes into account organisms of varying sizes and since we are concerned only with quantity we avoid having to recognize in-dividuals. We can also estimate abundance in terms of energy. We can find out how many calories there are in vegetation and in animals, not an easy job as special equipment is required, but one which is perfectly possible in a good ecological laboratory.

If our estimate is expressed in terms of dry weight per unit of area per unit of time we have a measure of abundance expressed in terms of productivity and this turns out to be the best in situations where we are trying to understand the transfer of energy in ecosystems.

In Chapter 4 we discussed food chains and food webs and constructed diagrams showing the possible feeding relations of the pike. Food chains and webs can now be re-examined in terms of energy transfer. Food chains often have three links, some have four or five, and a few go beyond this number. The various steps in a food chain are called trophic levels, a term which has much the same meaning as feeding levels, but is more inclusive as it incorporates the acquisition of energy by plants, a process which cannot be envisaged as 'feeding'. The various trophic levels in an ecosystem can be listed as follows:

*Producers*, normally photosynthetic plants which capture radiant energy from the sun and convert it to stored chemical energy.

*Primary consumers*, usually herbivorous animals feeding directly on green plants.

*Secondary consumers*, carnivorous animals commonly called predators or parasites and (if they are feeding on dead animals) scavengers. All feed on primary consumers.

*Tertiary consumers*, animals feeding on secondary consumers.

*Higher-order consumers*, all other predators, parasites, and scavengers feeding chiefly on tertiary consumers but occasionally at an even higher trophic level.

None of these distinctions should be interpreted too rigidly, except in the case of producers which can be defined with confidence as they have the special ability of being able to create organic matter by photosynthesis. Many consumers are opportunist and can switch between trophic levels as they feed. A bear feeding on fish and berries is both a tertiary and a primary consumer.

We must at this stage distinguish between consumers feeding on living organisms and those feeding on dead organisms. Land plants tend to be woody and contain in their tissues large quantities of cellulose which cannot be digested by most primary consumers. For example an elephant feeds continuously and during a day eats enormous quantities of plant material, but an examination of the quantity and quality of elephant faeces shows that most of the material eaten is passed straight through without being digested. The same applies to caterpillars feeding on the leaves of plants: large quantities are eaten and large quantities are defaecated.

Only a tiny fraction of what is available is utilized by primary consumers feeding on living land plants, the remainder is utilized by organisms in the same trophic level once the plants have died. There is no significant accumulation of dead plant material; all of it eventually

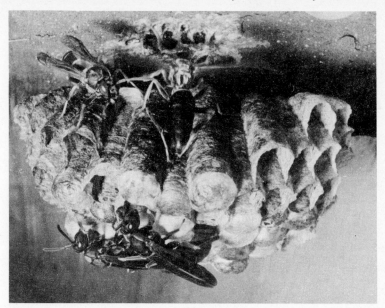

*Plate 15:* Nest of a North American wasp, *Polistes fuscatus*. One egg is laid in each cell and the resulting larva is fed by the adult wasps on the caterpillars of moths. The adult wasps feed themselves on nectar. Thus the adult is a secondary consumer when feeding its young and a primary consumer when feeding itself.

disappears, and the organisms responsible for its disappearance are called primary decomposers. They are in the same trophic level as primary consumers but differ from them in that they feed on dead rather than living vegetation. There are also secondary, tertiary, and higher-order decomposers (frequently called scavengers) feeding on dead consumers.

Returning to the pike, we are now in a position to construct another diagram showing part of the food web. This time energy loss occurs each time materials are transferred from one trophic level to the next. Because of its shape this type of diagram is often called an ecological pyramid; its most significant feature is that it is stepped and does not reach a point gradually. Fig. 19 shows a possible pyramid for the pike. The units could be of energy expressed in calories, of biomass expressed as dry weight, or of numbers in terms of individuals, although for reasons just given this last is the least satisfactory of the three possiblities. Fig. 19 suggests that there are four trophic levels centred on the pike, but there could be more: perch feeding on roach could be eaten by pike and this alone would add another level. The diagram omits other organisms feeding on the plants, and assumes that the pike is large and therefore the top predator in the ecosystem (for the moment we have ignored the angler). When we look at a pyramid of this sort we should bear in mind that the water providing the medium on which the ecosystem is based is not closed to other organisms: land animals and dead leaves fall into the water by accident and therefore contribute to the energy budget of the ecosystem.

Let us now consider a fully worked out example of an ecological pyramid. The ecosystem is a shallow pond and it was possible to obtain an overall estimate of productivity and biomass, and also the number of individuals of all the organisms in the pond. The pyramids of biomass and productivity are shown in Fig. 20. There is the expected big step between the producers and the primary consumers and thereafter the steps are smaller. The two pyramids are similar in shape and they are in fact providing more or less the same information in different units.

It is possible to construct other sorts of pyramids and pyramid

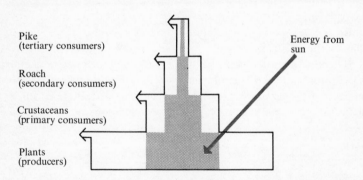

Pike
(tertiary consumers)

Energy from
sun

Roach
(secondary consumers)

Crustaceans
(primary consumers)

Plants
(producers)

*Fig. 19:* Part of a food web of a pike showing how energy is captured and transferred through trophic levels. The shaded area shows the energy retained by the system after each transfer and the arrows indicate energy loss. The diagram is very much a generalization, and shows only part of the energy relationships of the pike.

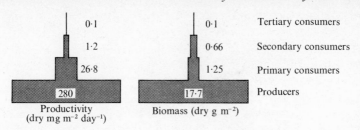

*Fig. 20:* Pyramids of productivity and biomass for a shallow pond. (Adapted from R. H. Whittaker, 1970)

building is an amusing and instructive ecological exercise. One easy project is to collect samples of insects from vegetation with a sweep net. A sweep net is a strong canvas bag attached to a stout wire frame and fixed to a short handle. The diameter of the net should be about 40 cm but it need not necessarily be circular; some people prefer a triangular shape. As the collector walks forward he swings the net to and fro as close to the ground as possible and after taking perhaps twenty-five sweeps the contents of the net are emptied into a tin containing a few drops of chloroform which immobilizes active insects. The material is then sorted, the vegetation removed, and the insects and other small animals like spiders are sorted into size classes, species, or by their known feeding habits. The latter two operations require a certain amount of entomological knowledge but even without this it is easy to classify the insects by size.

If the insects are sorted by length (a convenient measure of size) and a pyramid of size classes constructed it will normally show the smallest insects as the most abundant and the largest as the least abundant, but sometimes there are exceptions which should provoke us to ask questions. In the two pyramids shown in Fig. 21 the smallest insects are at the base, the largest at the top, and three additional size classes fit in between. The lower pyramid shows what we would normally expect and the upper one shows what was actually found in the sample. In this example an intermediate size class contained many more individuals than expected. This was because of exceptional numbers of one species, the meadow spittlebug, *Philaenus spumarius*, the insect whose immature stages produce the familiar 'cuckoo spit' on plants in the spring, and which is common in agricultural areas in much of North America and

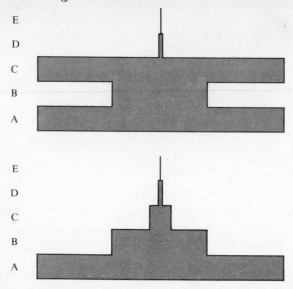

*Fig. 21:* Pyramids of numbers for insects collected by sweep netting in an abandoned field in Michigan. The upper figure shows the total sample and the lower figure after all individuals of one species, the meadow spittlebug, have been removed. The size classes (in mm) are: A, 0·6–2·5; B, 2·6–4·5; C, 4·6–6·5; D, 6·6–8·5; E, 8·6–10·5. (From F. C. Evans and U. N. Lanham, 1960)

Eurasia.* The lower pyramid in Fig. 21 shows the frequency of size classes when the meadow spittlebug is removed and it therefore appears that this single species is responsible for the distortion of the pyramid of numbers. We can therefore ask why the meadow spittlebug was so abundant in this sample. The collections were taken from an abandoned field in Michigan and the number of spittlebugs normally resident in the field is not especially high. But in July, when the sample upon which Fig. 21 is based was taken, many adult spittlebugs invaded the field from outside. They arrived after nearby clover and lucerne (alfalfa) fields were cut by local farmers. In North America meadow spittlebugs build up enormous populations on clover and lucerne and

*The cuckoo spit is produced by immature spittlebugs as they extract water and nutrients from plant tissues. Similar species which occur in Africa are known as 'snake spit' and there is as much superstition attached to these as there is to the meadow spittlebug.

when their environment is suddenly disturbed, as by cutting the crop, they take to the wing and move elsewhere. The result in this case is a sudden and unexpected distortion of the pyramid of numbers in another community, a good example of how human activities can disrupt an ecosystem. Indeed many human activities probably have similar effects on virtually all ecosystems but it requires the presence of an observant ecologist for them to be detected and described.

## Trophic models

A model in this context is an attempt to represent the working of the entire system. Most models combine assumption with fact and their function is to enable us to envisage how something works even though we lack certain essential information. Models must therefore be understood simply as ideas about how a process works and not as factual explanations of the process.

Various models have been proposed to suggest how ecosystems acquire and dispose of energy. None is entirely satisfactory. In 1942 R. L. Lindeman suggested a model for the transfer of energy in an ecosystem and his model has been adopted in one form or another by almost all subsequent ecologists. Lindeman's model, reproduced in Fig. 22, includes decomposers that receive energy from the producers and from all other trophic levels, but does not separate the decomposers themselves into trophic levels. But such a separation becomes necessary when we are dealing with decomposers of dead producers, as well as decomposers of dead animals, because these groups originate from different trophic levels. The Lindeman model does not allow the decomposers to be in different trophic levels, nor does it allow for the possibility of (for instance) a secondary consumer feeding on a primary decomposer.

Consider as an example a termite feeding on dead wood. It is clearly a primary decomposer and when it dies it might be eaten by the grub of a fly, a secondary decomposer, but it could also be eaten while still alive by an insectivorous bird, in this instance a secondary consumer. Indeed the trophic pathways of consumers and decomposers run parallel and this is allowed for in the modified model in Fig. 23. Here there are two pathways of energy transfer beyond the producers. The primary consumers utilize living producers and the primary decomposers utilize dead producers. Primary consumers are eaten by

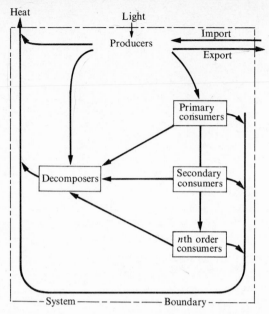

*Fig. 22:* R. L. Lindeman's model for the transfer of energy in an ecosystem.

secondary consumers or, if they die, by secondary decomposers which can also feed on dead primary decomposers. In Figs. 22 and 23 '*n*th order' refers to tertiary and higher-order consumers and decomposers and it is written like this simply to save space and repetition.

At each level there is energy loss through heat generated by work. Neither Fig. 22 nor Fig. 23 allows for species which as they feed move between trophic levels or are partly consumers and partly decomposers at the same level, but we are not so much concerned with species as with the trophic position of an individual whenever it acquires food. Moreover the models do not help in defining whether a resource is living or dead when it is utilized although in practice it is rarely difficult to decide whether an individual organism is taken dead or alive.

Man presents a special problem, as much of the food we eat has been 'killed' by other people. As a species we are essentially primary and secondary consumers. When we eat bread we act as a primary consumer, although the materials from which the bread is made have

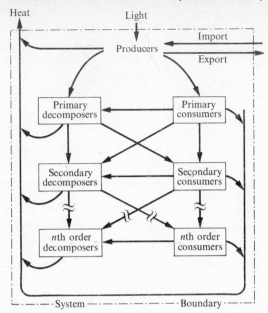

Heat  Light  Import  Producers  Export  Primary decomposers  Primary consumers  Secondary decomposers  Secondary consumers  *n*th order decomposers  *n*th order consumers  System ——— Boundary

*Fig. 23:* Another model for the transfer of energy in an ecosystem. (From R. G. Wiegert and D. F. Owen, 1971)

been 'killed' by other members of our species. When we eat beef we are secondary consumers, and with the exception of fish few of our foods place us in the tertiary or higher order trophic levels. We do not eat predatory animals to any great extent, nor do we feed on dead material, although of course we use vast quantities of 'dead' organic matter in the form of coal, oil, and natural gas, much of it being used to improve the productivity of the plants and animals on which we feed. Man is also unusual in the quite extraordinary geographical variation in diet: try fitting an Eskimo, an Indian peasant, and a European businessman into their respective trophic levels. We are without question the most difficult species to slot into trophic models simply because our sources of energy are so varied.

Fig. 23 clearly differentiates producers and the organisms that utilize them. Thereafter the levels can be placed in parallel provided we are able to distinguish between what is living and what is dead. How many trophic levels can we expect in an ecosystem? Most have at least three, some have four, but beyond this the number of organisms involved is so small that they play an unimportant part so far as energy transfer is concerned.

The structural complexity of terrestrial plants is a necessary consequence of positioning their photosynthetic parts so as to receive adequate sunlight. Air offers little support and the more conspicuous producers are large, woody, slow-growing trees and shrubs which because they take a long time to mature have low reproductive capacities. The bulk of the primary consumers in terrestrial ecosystems are insects which are small and numerous and have rapid rates of reproduction. There are of course many exceptions like elephants and antelopes. Primary consumers seem to have little impact on land plants – defoliation is unusual – and there is an abundance of living vegetation which is unexploited. But once the producers die they disappear through the actions of decomposers. In a forest ecosystem only a small fraction of the biomass is consumed while it is still living; the vast majority is decomposed once it is dead. This contrasts with an open-water ecosystem, such as a large lake or the ocean. Here the producers are supported by the medium (water) in which they live and do not therefore require the complexity of structure of land plants. Producers living in water are mostly small; they reproduce rapidly; and a high proportion, often approaching 100 per cent, is eaten by primary consumers. This means that the primary decomposer level is not well developed and most of the decomposers are secondary, feeding on dead primary consumers. The difference between an aquatic and a land ecosystem seems to result from the nature of the difference between water and air: water can physically support producers but land plants cannot be supported by air alone. In a mature deciduous forest only $1 \cdot 5 - 2 \cdot 5$ per cent of the net primary production is utilized by primary consumers, in pasture grazed by cattle the figure is $30 - 45$ per cent, and in most other terrestrial ecosystems the proportion lies between these two extremes. But in the open ocean, where the producers are floating plankton, between 60 and 99 per cent is utilized by primary consumers.

These differences between ecosystems based on land and water raise an important question about population regulation in different trophic levels. Earlier we have argued that organisms taken as a whole are limited by food as everything produced disappears and there is no accumulation. But in terrestrial ecosystems, and particularly in forests, the biomass of primary consumers is tiny relative to that of the producers, and hence the primary consumers make only a small impact on the producers. There is hardly any evidence of forests being badly damaged by primary consumers, although it may occur occasionally as when there is a sudden outbreak of defoliating caterpillars. On the other hand secondary and higher-order consumers appear to be limited by the numbers of primary consumers and frequently become short of food. Decomposers feed on all the organic material of dead producers and consumers and it therefore follows that as a group they are resource limited. In summary, all trophic levels beyond the producers are resource-limited, except perhaps the primary consumers on land, which seem to make little impact on their resources.

At this point we become even more theoretical. Consider first the aquatic (especially the ocean) ecosystem. The producers are mostly floating plankton, they reproduce rapidly and have short life spans, and there is a rapid turnover of populations. A high proportion is eaten by primary consumers, mostly small animals living near the surface of the water, and these in turn are utilized (when they are alive) by secondary consumers or (when they are dead) by secondary decomposers. This leads to a well-developed secondary consumer trophic level consisting mainly of small to medium-sized fish which in turn supports a conspicuous tertiary level consisting of larger predatory fish, many of them the familiar commercially important species. Thus in an aquatic ecosystem (and we are still thinking chiefly of the open ocean although the situation in large lakes is similar) there are four well-developed trophic levels made possible by the small size and edibility of the producers. The populations in each of the trophic levels beyond the producers are probably limited in size by available food. This does not mean that if we picked on a particular population we would necessarily find that its numbers are food-limited but it does suggest that taking each trophic level as a whole the availability of food is the most important limiting factor.

We seem to be facing a different state of affairs on land. Here the producers dominate the scene. Almost everywhere there is greenery and only occasionally do we find consumers having more than a slight effect on the biomass of plants. Primary decomposers are more in evidence than in aquatic ecosystems and although many are small and cannot easily be seen, the decomposition of dead vegetation is a familiar sight and the abundance of decomposers can be inferred, and of course if you look hard enough you will see them. Secondary consumers are relatively abundant and their numbers seem to be limited by the availability of primary consumers; there is rarely a well-developed tertiary level although there are of course individuals and species which are tertiary or higher-order consumers. There thus seems to be a fundamental difference between land and water, and initially it appears that primary consumers in land ecosystems occupy a unique position in that they have little or no effect on their food supply and therefore may not be food-limited.

This failure of the primary consumers on land to exploit fully the living producers can be interpreted in several ways. First it is possible, and many ecologists favour this view, that primary consumers are not food-limited but are limited by predators (which are of course secondary consumers). For example, we are asking in effect if the grazing antelopes of the African savanna are limited in numbers by their predators (lions, leopards, and other cats) or by the net primary production of grass. Or to take a more familiar example, are the numbers of caterpillars feeding on plants in our gardens limited by predation from birds or by the availability of plant material? There is evidence from the study of selected populations of primary consumers that predators play a key part in limiting numbers but this does not necessarily mean that primary consumers as a trophic level are predator-limited.

Land plants and the animals that feed on them have evolved side by side and tendencies by the animals to exploit more of the plant material are resisted by adaptations in the plants themselves. This is achieved in a variety of ways including the elaboration of thorns, tough leaves, and the presence of a quite extraordinary range of toxic compounds which play no part in the normal physiology of the plant but which deter animals from feeding on them. Some of our important insecticides like pyrethrum are plant products which have been evolved by the plants as natural insect repellents. All of this can be interpreted in the light of the need for land plants to be large and complex with much supporting tissue. Without these properties the medium in which they must photosynthesize would not support them, and there would be no terrestrial ecosystems as we know them. If the primary consumers never allowed land plants to develop their complex structures they would not themselves persist. The alternative is that land plants are consumed only to the point where serious break-

*Plate 16:* Flowers and seeds of an African milkweed, *Gomphocarpus*. Milkweeds contain heart poisons in their tissues which prevent or restrict attacks from herbivorous insects and other animals.

down and destruction of ecosystems is prevented.

We must remember that we are looking at something which has taken millions of years to develop and whatever the explanation it is clear that on land, though not in the sea, the greater part of net primary production goes to primary decomposers and not to primary consumers.

## Energy flow in and between ecosystems

Energy is not recycled because at each transfer from one trophic level to another much of it is changed into energy of a different kind which is no longer available or useful to organisms; the heat lost to the atmosphere as a by-product of animal and plant metabolism is in biological terms effectively useless. All energy transferred through an ecosystem is ultimately non-recoverable and in this respect contrasts with nutrients and other essential materials which are recycled; but we must appreciate that energy is expended to power the recycling of materials. Both the kind and the amount of energy available are vital to the maintenance of productive ecosytems. Oil and wind provide different kinds of energy and we therefore have to use them differently; similarly, an input of one kind of chemical energy can increase the productivity of an ecosystem, whereas the same amount (in terms of calories,* the units used to measure heating ability and hence energy content) of another kind would hinder not help: sewage effluent can cause coastal ecosystems based on phytoplankton to boom while an oil spill has the opposite effect, even though both are energy inputs.

The efficiency of energy flow within an ecosystem sets limits on photosynthesis, gross primary production, and the number of trophic levels. The energy output of an ecosystem, whether a forest or a field of cabbages, depends upon the energy input and the pathways through which the energy flows. Usually all energy is dissipated as heat, but it may remain locked up as chemical energy for a long time, as in the wood of living trees. About three hundred million years ago, vast quantities of energy over and above the apparent requirements of the ecosystem were locked up and eventually became coal; similarly, about one hundred and fifty million years ago the bodies of countless millions of small marine animals were not completely decomposed and

---

*Energy values in this section are given in kilocalories or Cal (1000 calories), the units used to describe the energy content of human food. The joule is increasingly used as the basic energy unit: 1 joule = 0·239 of a calorie.

eventually became oil. Coal and oil are thus fossil energy of biological origin.

Consider a hedgerow as an ecosystem; throw an apple core into it and you increase the energy input which may contribute in a small way to the bound chemical energy of blackberries produced later in the year. Walk away with the blackberries you pick later on, eat them, and dispose of the indigestible wastes elsewhere: you have removed energy from one ecosystem and placed it in others, using your own energy to do so. In autumn watch a squirrel burying a hazlenut or a jay burying an acorn and you witness animals locking up energy that can be used later. All around us chemical energy is being moved from its place of origin and eventually released elsewhere as heat energy.

Solar energy is effectively the only energy input to certain ecosystems such as the ocean and large lakes. These are sometimes called solar-powered ecosystems as they depend only on the radiant energy that penetrates the earth's atmosphere which at different latitudes varies between one and two million Cal to each square metre during a year. The energy flow through solar-powered ecosystems is determined simply by the efficiency of energy conversion; their annual energy flow achieves values of between one thousand and ten thousand Cals per square metre per year, which is minute compared to what a modern power station can produce. Many ecosystems receive a natural energy subsidy which augments the radiant energy captured by plants and as a consequence energy flow is substantially increased, often by a factor of ten or more. Tidal action in an estuary, dead leaves falling into a pond, heavy rain on a tropical forest, and even aphids from the countryside settling in gardens all produce energy subsidies which make more nutrients available and hence increase production.

Man-subsidized ecosystems include all kinds of agriculture, silviculture, and aquaculture with products as diverse as wheat, battery hens, timber, and carp. The energy subsidy may take many forms: irrigation, fertilizer, transport, weed removal, pesticides, and so on. These ecosystems rely upon energy diverted from one place to another or from one time to another; in modern farming the use of fossil fuel in the form of oil is by far the most important kind of subsidy. Originally man's subsidy was his own labour, and this is still true in fields cultivated by poor people who lack machinery and oil, and in gardens tended by the more affluent. Energy subsidies for agriculture increase crop yields but costs may limit the usefulness of these subsidies. For example, the production of potatoes in Britain in 1970 used as much energy as was contained in the harvested crop. Potato-growing is profitable because at the moment the cash value of the energy subsidy is less than that of the crop: in ecological terms there is no difference, the Cals put in equal the Cals taken out. As fossil fuel sources are depleted the use of oil as a subsidy to agriculture will become more and more expensive and ultimately will be prohibitive. And while there may be large reserves of oil locked up in shales, their value lessens as the cost of extraction approaches the value of the oil recovered.

Our main contribution to energy flow between ecosystems is in transporting energy from one place to another on a scale that never occurs in nature. The energy flow in a modern industrial city is enormous, but there is little primary production and most of the input is in the form of fuel and materials from elsewhere. A city is dependent on massive subsidies from other ecosystems and requires many times its own area to support itself. Are we justified in calling it an ecosystem at all? Probably not, even though it possesses many attributes characteristic of ecosystems.

## Cycles of essential materials

There is more to ecosystems than the transfer of energy between trophic levels. Organisms require other materials, water being the most obvious; they need nitrogen to make essential protein, and carbon in the form of carbon dioxide without which photosynthesis would be impossible. There is also a variety of inorganic materials required in small quantities, including the metal salts that are taken up in solution by plant roots, or in the case of aquatic plants directly from the water in which they live. Animals acquire these materials when they eat plants or other animals and so in this section our interest is centred on how plants obtain essential materials.

All of the forty or so chemical elements present in organisms occur in the non-living part of the earth's crust, but in different proportions: organisms contain relatively more oxygen, carbon, hydrogen, and nitrogen than non-living (inorganic) matter, but less potassium, sodium, iron, and silicon. These materials circulate through ecosystems, some as elements and others as more complex chemical compounds. All are important in one way or another for the maintenance of life. Unlike energy, which is lost from ecosystems, the materials under consideration are cycled again and again. The cycles

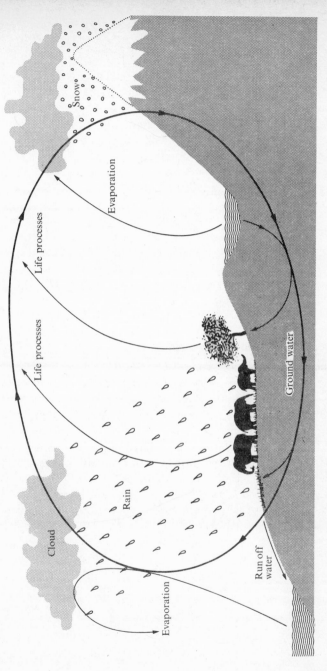

*Fig. 24:* The water cycle. There are really two cycles, one involving organisms and the other not.

for the various elements are well understood but we cannot consider all of them and will instead examine some of the more important.

Water makes up about seventy per cent of the weight of most plants and animals, including man. Aquatic plants clearly have no difficulty in acquiring water. Land plants absorb it through their roots, but the soil may become dry if there is little rain and under these conditions growth is inhibited; indeed the availability of water determines the distribution and abundance of many species of land plants. The land's water comes from rain (sometimes falling as snow or hail) and much of the water received returns directly to the atmosphere through evaporation. In the tropics the evaporation rate is high and even in areas of moderate or heavy rainfall there is not always enough water to sustain the growth of some plants. In temperate areas the evaporation rate of water from land is much lower and even light rainfall can be biologically more effective than two or three times as much in hot tropical areas. Evaporation also occurs from the surface of the sea and from fresh waters. Water from rainfall is taken up by land plants but relatively small quantities are retained; most is lost through transpiration and returned to the atmosphere. Animals too lose water through evaporation and in their waste products. There are, then, two water cycles, one through organisms and one which by-passes them, as shown in Fig. 24.

As already explained the process of photosynthesis makes use of carbon dioxide, which is converted into organic compounds and stored, eventually to be utilized by consumers and decomposers. All organisms release carbon dioxide into air or water during respiration and during decomposition of their waste products, and (when they die) of their bodies. Most of the carbon dioxide released is recycled and there is, as shown in Fig. 25, an effectively closed system.

The nitrogen cycle is more complicated. The earth's atmosphere contains seventy per cent nitrogen gas and although nitrogen compounds are essential to organisms for the formation of protein, few can make direct use of the gas from the atmosphere. Animals obtain protein from their food; as is well known some foods contain more protein than others, and a shortage of protein leads to malnutrition. Protein malnutrition seems unusual in wild animals but it is common in man because there is a world shortage of food rich in protein and because certain groups of people (for reasons not fully understood) are prejudiced against eating protein-rich food.

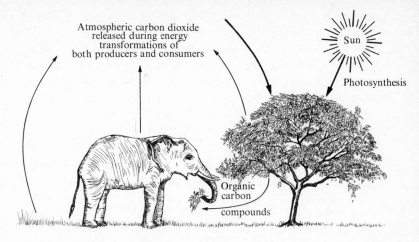

*Fig. 25:* The carbon cycle. Carbon in the form of carbon dioxide is obtained by producers from the atmosphere or from water. It is returned to the environment by the respiration of organisms of all trophic levels. The elephant is a primary consumer feeding on grasses and on trees.

Fig. 26 shows the nitrogen cycle. There is really more than one cycle and we could begin anywhere in describing it. If we start with dead plants and animals it can be shown that as they decompose simpler compounds are formed, including ammonia, which being soluble in water can be taken up by the roots of plants and returned after decomposition of the plants, or of the animals that feed on them. Some

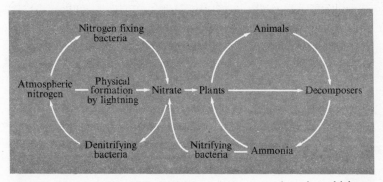

*Fig. 26:* The nitrogen cycle. There are really several cycles which are explained in the text.

of the ammonia goes through a different cycle. In the soil there are two groups of bacteria which convert ammonia into nitrites or nitrates. Nitrites cannot be used by most plants, but nitrates can and are an important source of the nitrogen required for the formation of protein. The soil bacteria which convert ammonia into nitrites and nitrates depend on a source of oxygen, and if the soil is deficient in oxygen because it has been flooded they cannot function. Under these conditions another group of bacteria convert nitrogen-containing compounds to nitrogen gas which passes into the atmosphere where, were it not for two further processes, it would leave the cycle. The most important of these processes involves yet another group of bacteria which are able to convert atmospheric nitrogen into nitrates, and thus bring the nitrogen back into circulation. Some of these bacteria are free-living, but one group occurs in roots, especially of plants of the bean family. The bacteria cause distinct swellings in roots which can be seen easily with the naked eye. If you pull up bean plants you can often see these swellings in the roots and mistake them for some kind of disease. Farmers know that if the soil is poor in nutrients and unproductive it is a good plan to grow a crop of plants of the bean family for a season. In temperate areas crops like clover serve this purpose but a crop of edible beans or peas is just as good. The plants enrich the soil through the action of the bacteria in their roots, so that soil fertility is restored, and a more desirable crop can confidently be grown next season. Crop rotation of this kind has long been farming practice in many parts of the world, and it simply acknowledges that periodically the soil requires a boost of nitrogen. Much the same effect can be produced by the application of nitrogen-rich fertilizers.

The remaining part of the nitrogen cycle shown in Fig. 26 is perhaps of more common occurrence in the tropics than elsewhere. It is entirely physical and occurs during thunderstorms when lightning converts atmospheric nitrogen directly into nitrogen compounds which can be utilized by plants, but its importance on a world scale is not known.

Most elements pass from one organism to another without ever entering the atmosphere. Many sorts of rock contain calcium compounds some of which are soluble and consequently occur in water. Plants take up calcium in solution through their roots; animals acquire it by drinking or eating; and it is returned to water or soil during the decomposition of dead organisms. Calcium is incorporated into skeletons and shells as calcium carbonate. After death shells in particular tend to accumulate at the bottom of ponds, lakes, and the sea and eventually become compacted into rock. Major earth movements result in marine rocks being uplifted to form dry land or even mountains. Rain as well as ground and soil water then dissolve calcium compounds from the rocks and the cycle repeats itself. The release of calcium from rocks is closely linked with the release of carbon since calcium usually occurs as calcium carbonate. In areas deficient in calcium, snails eat the shells of dead snails. Some species, including *Afropomus balanoidea*, which is found in acid streams in forests in Sierra Leone and Liberia, remove the apical whorls of their own shells and re-deposit the calcium at the growing lip of the shell.

Plants take up sulphur as sulphates in solution and incorporate it into proteins. Decomposition of plant or animal remains and animal excretion result in the release of hydrogen sulphide gas and the return of sulphates to the soil or water. Two sorts of specialized bacteria play major roles in the sulphur cycle. One converts some of the hydrogen sulphide to sulphates while the other releases hydrogen sulphide from sulphates. Under anaerobic conditions where there is a lack of oxygen much of the hydrogen sulphide passes into a reservoir of deep sediments or undisturbed, water-logged soil where it is slowly converted into other compounds and released again.

The numerous other cycles of inorganic materials follow a variety of patterns and the essential point is that the materials are circulated again and again. Once again an exception must be made for man. We not only cycle materials (some of them harmful pollutants) but also lock up more or less permanently large quantities in our buildings and machinery and, perhaps more seriously, we tend to disperse materials after they have been used in such a way that their future recovery is virtually impossible.

## Ecosystems and food production

Very few people in the world still hunt for wild animals and gather wild fruits and vegetables as their chief source of food. The bushmen of the Kalahari Desert and the forest-dwelling pygmies of the Congo Basin in Africa are among the few that subsist by hunting and gathering, but their numbers are declining and they are rapidly adopting the livestock-rearing and crop-growing technologies of neighbouring tribes.

Nowadays most people live in societies which practise some sort of agriculture in which energy flow and the cycling of materials is controlled in order to produce food. Even so, more than a thousand million people suffer some degree of malnutrition; their stomachs may be full but their diet lacks vital nutrients or contains nutrients in the wrong proportions; 700 million people do not get enough to eat. A widespread form of malnutrition results from lack of protein in the diet; proteins are essential for growth, and consequently it is the children that suffer most. Malnutrition can also be caused by too little carbohydrate in the diet so that protein needs to be utilized to supply energy. Overconsumption of fat and carbohydrate leading to obesity is another form of malnutrition. Then there are numerous disorders resulting from vitamin deficiencies. But protein-energy malnutrition is the most serious, affecting a vast number of infants and children in the developing world.

Can the world population be adequately fed? In 1967 man produced more than enough food to give each person his minimum annual requirement of one million Cals, but inequalities of distribution meant that some had too much and many too little. This harvest represents less than one per cent of global net primary production, implying that more food could and should be produced; but there are reasons why it may never be substantially increased. Much of the available fertile land is already under cultivation, and extending agriculture would require massive energy subsidies and would put water recycling at risk. A solution to feeding an expanding population adequately would be to make better use of the available ecological information about trophic levels, energy flow, and the cycling of nutrients in different kinds of ecosystems. In pasture only 30 – 45 per cent of net primary production is utilized by cattle; people cannot eat the leaves and stems of grass (a cow's 4-chambered stomach houses a symbiotic flora and fauna of bacteria which facilitate grass digestion) but they can eat grass seeds (cereals), potatoes, and beans, which supply all the required protein. A hectare of land that can yield 370 kg of potato protein yields only 110 kg of beef protein, and it is thus more efficient to grow potatoes than to raise cattle. This is not to suggest that we all become vegetarian — in most countries there are certain areas that only livestock can use — but a diet of wheat, beans, and potatoes, with a little meat, provides the right sorts of proteins for health and growth. Some believe that Britain could support five times its present population without importing food (nearly fifty per cent is currently imported) if instead of feeding grain and other plant material to livestock we became vegetarian and ate nothing but home-grown crops.

Maize is the third most important world crop and is about average among temperate region crops in the amount of energy subsidy needed to grow it on a commercial scale. It is calculated that 1·26 per cent of the solar energy reaching a maize field is converted into maize plants while only 0·4 per cent ends up as stored chemical energy in the grain. In 1970 farmers in the United States subsidized commercially grown maize with fossil fuel equivalent in energy to eleven per cent of the solar energy captured by the plants. This seems a small subsidy, but fossil fuels are a finite resource and sunlight is unlimited. Between 1950 and 1970 the yield of maize in the United States more than doubled, but the ratio of energy output (contained in the maize) to input (the energy subsidy) fell by eleven per cent. During this twenty-year period manpower used to grow maize was halved, but the inputs of oil, fertilizer, insecticide, weed-killer, electricity, and transportation all increased substantially. The farmers continued to make a profit but at enormous ecological cost: for how long can we expect modern farming to exist in this way?

, Food produced with the help of large energy subsidies is expensive. Although much of the world population depends on grain crops, the poorest and most needy countries cannot afford to purchase the surplus produced by the rich countries. The problem of feeding the under-nourished world is as much a problem of distribution and economics as of agricultural productivity. The European Common Market countries will continue to produce 'mountains' of certain foods, but there seems little that the ordinary citizen can do about this.

## Ecosystems and man

This chapter has often drawn our attention to man's peculiar role in world ecosystems; indeed all the varied events we have been speaking about can potentially be altered and disrupted by human activities. We modify natural ecosystems and in doing so frequently create new ones. Nature could never produce something as bizarre as a field dominated by cabbages one year and clover the next, and we should therefore not be astonished to learn that human activities are having profound effects on virtually all ecosystems. The use of artificial fertilizers, pesticides, and farm machinery has made possible novel

*Plate 17:* Mount Kenya in East Africa. The communities on the slopes of high tropical mountains are among the few that have remained undisturbed by man, but even here there is a risk of air-borne pollution.

ecosystems whose essential function is aimed at producing more and more food for more and more people. Agriculture, industry, and many of our leisure activities (like motoring) result in the pollution of the land and air, and it is probably true to say that almost every lake and river has at least to some extent been polluted. Even the open ocean and the polar ice caps have not escaped, and the only places likely to be free of human pollution are the tops of high mountains.

The damaging effects of pollution often occur at a considerable distance from the place where the substance was first applied, and sometimes a long time afterwards. In some places, Illinois for example, nitrates applied as fertilizer are percolating into drinking water supplies. In twenty years time the underground water which supplies some thirty per cent of Britain's needs could contain concentrations of nitrates which would damage the oxygen-carrying capacity of the blood of small children. Sufficient phosphorus (also from fertilizers) has accumulated in the sea off Newfoundland to tinge red the flesh of herrings. A special problem arises with synthetic substances that plants and animals cannot break down. These substances not only persist but become concentrated up the food chain, so that a top predator contains many times the amount found in a producer or primary consumer. D.D.T. and other pesticides are persistent poisons that are virtually indestructible; water, soils, and organisms somewhere in the world probably still contain all the D.D.T. that has ever been applied. The rich industrial nations that manufacture pesticides are increasingly aware of the dangers of persistence and accumulation, but double standards prevail: pesticides whose use at home is now banned are manufactured and exported to developing countries by the United States and several western European nations. Many of the importing countries have no regulations concerning the use of these poisons and the ordinary agricultural worker is ignorant of the dangers to his own health that can result from careless handling. Ironically, some of the exported poisons are eventually imported in food sold by the poor to the rich countries. In terms of pollution, paper production is one of the dirtiest of all manufacturing processes; in Scandinavia, Finland, and Canada effluents from pulp mills have reduced the oxygen content of rivers and lakes and poisoned aquatic ecosystems with mercury, thereby damaging water supplies, fisheries, and recreational facilities. In 1978 animals were killed and people taken ill by poisonous vapours blowing from the sea onto the coasts of southern Brazil and Uruguay. The Brazilian Ministry of Health attributed this to fumes from dead algae but it now appears that a ship carrying dangerous chemicals, believed to include mercury, was wrecked seven years earlier on a reef off the coast of Uruguay and broke up in storms in late 1977 or early 1978. The local people feared poisoning from contaminated water and fish but faced a conspiracy of silence over the real nature of the cargo and the dangers it posed. Awareness of the dangers of pollution is not new. In the Middle Ages the inhabitants of upland Derbyshire mined lead and also farmed on a small scale. They planted trees along the mine workings and around spoil heaps to shade the ground, which discouraged the growth of grass and thus reduced the consumption of contaminated grass by cattle.

All of man's activities result in increases and decreases of other species which, as we have seen, are likely to alter the functioning of ecosystems. We are only just beginning to realize the magnitude and irreversibility of many of these changes. The last chapter of the book discusses the implications of human activities for the natural world, and here we end with just one example that we can explore using the knowledge gained from the discussions in this chapter.

Let us consider the different consequences of cutting down a temperate forest and a rain forest. We shall neglect the differences in species diversity between the two ecosystems and the effect of upsetting this diversity, and concentrate on a single event. Fig. 27 shows the distribution of organic carbon in a northern coniferous forest and a tropical rain forest. Both ecosystems contain about the same amount of organic carbon. In the coniferous forest about half is in the soil while in the rain forest about three quarters is in the wood. This difference results from the fact that in a temperate forest decomposition is slow, presumably owing to the absence of termites and other important decomposers, and perhaps also because of lower temperatures, while in rain forest it is extremely rapid and most of the organic carbon (and

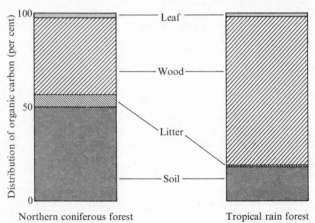

*Fig. 27:* Distribution of organic carbon in a northern coniferous forest and a tropical rain forest. (From E. P. Odum, 1971)

indeed other material) is locked up in living vegetation rather than in the soil. Similarly about fifty-eight per cent of the nitrogen in a rain-forest ecosystem is in vegetation while in a British pine wood the corresponding figure is less than ten per cent. Thus between tropical and temperate forest ecosystems there is a striking difference with far-reaching implications. If a rain forest is cut down and the timber removed, most of the nutrients are also removed, but if temperate forest is cut down a substantial proportion of the nutrients remains in the soil. In temperate areas the soil left after the removal of timber is still quite fertile, but where rain forest timber is removed the remaining soil is almost devoid of nutrients. This explains failures when cultivation is attempted in tropical areas cleared of trees. The situation is further aggravated in rain forest because inorganic cycling tends to occur rapidly, and heavy rainfall results in soil erosion and leaching of nutrients. This example illustrates one of the many effects of human disturbance of natural ecosystems.

# 6 Natural selection and adaptation

The ability of a species to fit into an ecosystem and to persist generation after generation depends largely on the quality of the individuals that make up the species. By quality we mean the genetic constitution of the individual and its ability to transmit useful attributes to its offspring. The genetic make-up of a species is ultimately determined by the environment, and therefore by the ecosystem, in which it occurs. The environment influences which individuals survive, and which die, through a process of non-random elimination called natural selection which we shall examine in more detail a little later in this chapter.

The inherited characteristics that enable individuals to survive constitute adaptations to the environment. Adaptations are distinct from adjustments to the environment made during an individual's lifetime. Thus plants and animals may adjust to the nutrient supply by attaining maturity at different sizes: cabbages grown in poor soil are smaller than usual and starved caterpillars turn into under-sized butterflies; but providing the offspring receive adequate nutrients they are of normal size. We can therefore draw a distinction between genetically determined adaptation and environmentally determined adjustment, although in practice it is often difficult to separate the two.

Climax communities tend to persist in a state of equilibrium year after year, the species of plants and animals occurring in much the same numbers, some common and many rare. There are of course fluctuations in numbers of some species but these are small compared with what is possible. Imagine a species entering an ecosystem for the first time. The individuals involved are unlikely to be well-adapted to the special qualities of their new ecosystem and their chances of survival and therefore of the species persisting through successive generations are slight. Ecosystems supporting climax communities are remarkably resistant to colonization by new individuals and usually it is only when there is disruption of the long-established equilibrium that additional species can become incorporated. But once a species has penetrated an ecosystem it may in time become better adapted to the new environment. It is the ability to adapt to changed circumstances that we shall be mainly concerned with in this chapter.

The physical features of the surface of the world have changed enormously since life began, and the configuration of continents and oceans has been quite different in the past from what it is now. The presence of fossils of marine animals in places now far from the sea is but one piece of evidence supporting this statement. There have also been great climatic changes, and the world has been much colder and much warmer that it is today. Even as recently as 10,000 years ago it was much cooler in the northern hemisphere than it is today: the Sahara Desert which nowadays supports little life was covered with vegetation, and some of the larger mammals now confined to East, Central, and South Africa once lived there. Areas of dry savanna were once tropical rain forest, and during the past million years the extent of rain forest in Africa has been both smaller and larger than it is now. Thus despite the apparent stability of ecosystems with climax communities, we must realize that there has been considerable long-term change, providing scope for species to adapt, and there is every reason to suppose that these changes will continue. Short-term changes can be seen when species establish themselves in ecosystems for the first time, when they disappear, and when rare species become common and common species rare. All manner of factors can be invoked as generating change, but climate is probably the most important, except of course in the last 10,000 years or so when man has cultivated the earth and so made the most dramatic changes to the natural world. Man-made changes in natural ecosystems have occurred and are still occurring at a rate far greater than those brought about by natural events. A few hundred years ago southern England was covered with oak forest; now the forest has almost gone, and the landscape consists of neat little fields surrounded by hedges, towns and villages, and vast urban sprawls — an entirely different environment. The environment of modern England supports ecosystems very different from those of past centuries: it has been colonized by species more characteristic of open country and some of the plants and animals associated with the oak forests have become rare or have disappeared.

Over the last twenty-five years the Forestry Commission has planted thousands of hectares of northern Scotland with the lodgepole pine, a fast-growing tree first introduced from North America in 1854. The tree thrives on deep peat in exposed sites where the native Scots pine grows less well. Caterpillars of the pine beauty moth, *Panolis flammea*, feed on the leaves of Scots pine but rarely cause damage. However, in 1977 pine beauty caterpillars destroyed about 240 hectares of lodgepole pine, and by June 1978 numbers were such that a newspaper headline designated them as 'a moth pest worse than elm disease'. Up to four hundred pupae per square metre were found beneath defoliated trees, an extraordinary density quite uncharacteristic of the species, but known to have occurred before in plantations of in-troduced pines elsewhere in western Europe. The moth has apparently changed to exploit a new food source, but it is not known whether the change is an adjustment or an adaptation or a combination of the two.

## Natural selection

The theory of natural selection, the most unifying of all biological theories, was first proposed by Charles Darwin in the middle of the nineteenth century as a way of explaining evolution and the descent of plants and animals from more primitive forms. It depends on a series of propositions, all of which can be observed and tested. The first of these we have repeatedly stressed in this book: organisms possess the capacity for exponential growth in numbers, but despite this most populations do not grow but remain relatively stable – vast numbers of individuals die before they reproduce. The second is that organisms are individually variable – no two are exactly alike – and the majority of individual characteristics are inherited. Third it can be deduced from these observations that survivors are likely to differ in certain respects from those that die, which is another way of saying that the probability of death varies between individuals and depends on their genetic make-up. Here we need not specify the nature of individual characteristics except that they include size, proportions, colour, and pattern. A glance at a crowd of people in the street will confirm that no two are exactly alike, and most of us also know of 'invisible' differences like the blood groups.

An individual's probability of surviving and reproducing is called its fitness. Each individual because of its genetic constitution will have a certain probability of growing, escaping enemies and the hazards of weather, and of enjoying food, light, and other resources. The term 'survival of the fittest' refers to this probability.

Natural selection or survival of the fittest must be viewed as a statistical process in which some individuals have a better chance of surviving and leaving offspring than others. Under most circumstances natural selection imparts genetic stability to a population; the same kinds of individuals are the survivors generation after generation, the population remains unaltered, and there are no adaptive changes. But if the environment alters, a different set of individuals with different genetic traits may stand a statistically better chance of surviving and reproducing. Given time natural selection alters some of the characteristics of the species so that it is better adapted to the changed environment. This kind of adaptive change is the equivalent of evolution, a phenomenon which in its simplest terms is net genetic change. Species are constantly subjected to the possibilities of alteration but the process of evolution is usually slow and we should not normally expect to be able to see adaptations developing.

We have so far been speaking of natural selection as acting on the existing variation among individuals of a population and this is what usually happens, but the genes determining the appearance of an individual are also liable to mutation, an event which provides a supply of new variation in the population. Mutations occur infrequently, often no more than once in every ten thousand to a million individuals, and sometimes even more infrequently. We can view mutations as haphazard changes in the chemistry of the genes, normally disadvantageous, with the result that affected individuals or cells or foetuses are eliminated by natural selection. Occasionally an advantageous mutation occurs and if environmental conditions are right it will quickly spread through a population. It cannot easily spread by itself; the environment has to favour the mutant individuals to such an extent that their chances of survival are greater than those not affected. Since mutations occur infrequently we can deduce that the probability of one occurring will depend on the size of the population. A small population of a thousand individuals with one generation a year may produce a particular mutation only once in ten years, but a larger population of the same species containing 10,000 individuals in each generation will on the average produce the mutation once a year. Mutations are not in themselves important in the process of adaptation but they are the main source of new variations that provide the raw material upon which natural selection can act.

Ecologists cannot afford to neglect natural selection and the capacity of organisms to adapt to new circumstances. We as people are responsible for the most conspicuous environmental changes now taking place on earth and in searching for examples of natural selection inducing adaptive changes in populations we should look first at environments we have modified and created.

## Natural selection and evolution

Darwin appreciated that natural selection usually produces genetic or adaptive stability. Only exceptionally will it result in change – that is evolution. Organisms less suited to a particular environment (including those with new mutations) tend to be eliminated, and an evolutionary change can in a sense be envisaged as a population's last resort, an attempt to cope with the effects of a changing environment, that occurs only when all other possibilities have failed. Since man is now the most important single factor initiating environmental change we should expect to find evidence of man deliberately or accidentally inducing evolutionary modifications in other organisms, and indeed in himself. The most obvious examples are in domesticated plants and animals which have been selectively bred for food and, in the case of dogs, cats, and canaries, for companionship. Domestication involving deliberate selective modification has been going on for at least 10,000 years. The process is usually known as artificial selection but it is essentially the same as natural selection and differs only in that by a process of trial and error man has purposely bred varieties of plants and animals considered desirable. Existing genetic variation has been exploited to breed thousands of different varieties of crops, useful new mutants have been carefully conserved, and species which would not otherwise do so have been forced to hybridize. Man has thus created an enormous diversity of organisms which would not otherwise exist: virtually all the varieties of plants in your garden are the product of man's ingenuity.

Crops are derived from wild species of plants but the ancestors of some of our common crops have evidently disappeared. Almost all our crops were discovered and selectively bred thousands of years ago and it is remarkable that with our recently acquired knowledge and technology we have been unable to add significantly to the range of useful species of crops. Some crops are now completely different from existing wild species and from every point of view they may be regarded as species created by man. One of the best known is maize (corn) whose ancestor, or ancestors, is believed to have come from South America. The maize plant is totally different from any known wild species; it cannot propagate itself, as the seeds must be released from the cob before they germinate, and it is therefore entirely dependent on man for its survival. It has been spread all over the world and is a staple crop in many tropical countries. There are hundreds of different genetic varieties, each selectively bred by man to suit particular environmental conditions; it is in every way a species created and diversified by man.

By breeding varieties of crops man has accidentally caused changes in the genetic make-up of species of weeds. Some weeds are especially associated with certain crops and cannot persist in other environments. Man is also responsible for all the breeds of cattle, sheep, pigs, goats, chickens, and other animals domesticated for food, and for dogs, cats, and other pets. Wild ancestors of these animals either no longer exist or are rare. With domestic animals as with crops man is responsible for much evolutionary change brought about by trial and error selective process, and there seems little doubt that as techniques and knowledge expand there will be an even greater proliferation of desirable varieties in the future, although strangely enough there seems little prospect of discovery of additional species suitable for domestication.

There are other perhaps more unexpected evolutionary changes brought about by man and as we search deeper more and more examples are coming to light; indeed we have probably under-estimated our powers of accidentally inducing evolution. Thus when synthetic insecticides were invented no one seriously entertained the possibility that insects would develop genetic strains conferring resistance which renders the insecticides ineffective. Attempts now-adays to eradicate mosquitoes are frequently frustrated because the mosquito populations adapt to the strong environmental pressure of a massive dose of insecticide. The evolution of insecticide resistance occurs by a selective process and after a time a whole population may become immune. Not all insect pests become resistant but it is anyone's guess as to what will happen in the future. Pest controllers and pesticide manufacturers are already facing a predicament which is effectively a race between the development of a new compound and the evolution of a new resistant strain of the pest. There is no obvious end to this race and we can do no more than simply try and keep up with nature. The difficulty is that pests tend to occur in huge numbers and breed rapidly, qualities greatly facilitating evolution by natural selection.

A parallel state of affairs exists in attempts to combat human diseases with antibiotics and drugs. It is becoming increasingly evident that many of the micro-organisms causing human diseases develop resistance; even penicillin is less useful than it once was as a means of controlling bacterial infections. From what we know of the capacity of organisms for growth in numbers, of the exceedingly high death rates, and the extraordinary amount of genetic variation present in most populations, we should not be astonished to discover that attempts to control and eradicate various micro-organisms responsible for human disease lead through selection to new strains against which our efforts are futile. Add to this the fact that the human population (which provides the environment for disease organisms) is not only large but is

*Plate 18:*  Spraying an East African swamp in order to destroy mosquito larvae. The product being used is made from oil.

increasing rapidly, and therefore provides a splendid arena in which disease organisms can adapt as a result of our rather clumsy attempts to control them.

We should also look for examples of evolutionary change among plants and animals deliberately or accidentally introduced by man into new environments. We can predict with confidence that if samples of plants or animals from areas where they have been introduced are examined and compared with similar samples from areas from which they were derived, differences will be found in the appearance of specimens. House sparrows from the middle of England were introduced into North America in about 1852 and eventually spread over almost the entire continent. Samples of sparrows obtained in the 1960s from many parts of North America indicate that since the introduction the birds have changed in several characteristics (such as size and proportions) and have become adapted to the local environments in which they now occur.

It is tempting to ascribe all apparently adaptive modifications to the effects of natural selection acting through pressures exerted by the environment, but there is more to it than this. If a population is started by only a few individuals it is unlikely to possess the full complement of genetic variation of the species. This means that at least initially a new population is restricted in evolutionary potential; natural selection cannot act without the appropriate genetic variation and only later as favourable mutations occur is it possible for the full adaptive prospects of the population to be realized. This may take time and it would therefore seem likely that many new populations are unable to exploit fully the environment in which they start because of the absence of part of the genetic variation of the species. The phenomenon we are speaking about is known as the 'founder effect' and it simply means that the subsequent genetic structure of a population depends in part on the genes present in the individuals that started the population in the first place. Biologists disagree over the extent to which the founder effect is important in evolution and adaptation, but there seems little doubt that its consequences are more marked in some environments than in others. Evidence for the founder effect is mostly indirect – it is one of those ecological phenomena where common sense would suggest that it must be happening but where it is virtually impossible to be sure.

Land snails are one group of animals in which the genetic make-up of the founders is likely to determine the subsequent appearance of the population. They are especially liable to accidental introduction into new areas and yet they are highly sedentary with limited abilities to disperse by their own efforts. Moreover some species are strikingly variable in conspicuous features like the colour and pattern of the shell and they have therefore attracted the attention of ecologists interested in the genetic aspects of the subject. We can take as an example one of the African species, *Limicolaria martensiana*. In Uganda it is associated with environments disturbed by man and like many snails it flourishes in gardens. It is enormously variable in colour and pattern but two

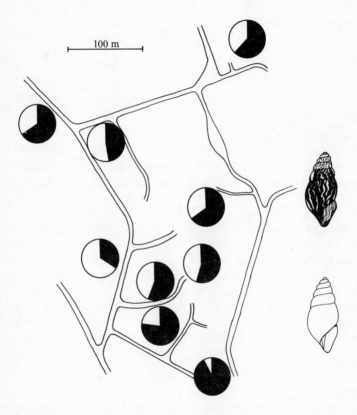

100 m

*Fig. 28:* Frequency of streaked individuals (black) and unstreaked (white) in populations of the snail, *Limicolaria martensiana*, in gardens near Kampala, Uganda.

genetically determined forms occur in most populations: one streaked with dark lines on a pale background, the other unstreaked. Fig. 28 shows the frequency of these two forms in populations in gardens near Kampala. As shown the frequency of streaked individuals varies between gardens and there is no obvious trend which would be expected if natural selection determined the adjustment of the two forms to local environments.

As everywhere else, gardeners in Kampala continually transport plants from one garden to another and new plants are often brought in from other places. The eggs and young snails are sometimes found attached to the roots of plants moved by gardeners and it seems likely that this is how the populations are started in the first place. All the populations shown in Fig. 28 are isolated from one another by ecological barriers like roads and open spaces and the only contact populations have with each other is when individuals are accidentally transported by gardeners. In this example man has created the barriers that separate the populations and it is probable that the differences in frequency of streaked and unstreaked snails in the gardens are the chance result of which kinds of snail were introduced into the gardens in the first place. This does not rule out the operation of natural selection in adjusting the frequency of the forms to the local environment and indeed it is not possible to state with confidence whether selection or the founder effect is the more important in this instance.

A parallel situation occurs when a cultivated plant like a marigold is introduced into a garden. Marigolds and many other garden plants have a variety of genetically determined flower colours and if they are allowed to seed themselves through several generations the colours of the flowers of the plants established in the garden will depend on the genetic make-up of the founders, although here again we cannot discount the possibility of natural selection playing a part as well.

Competition for resources between individuals of a species is an essential part of the selective process that leads to adaptation; competition between species leads to selection for differences in the way resources are exploited. Herbivores can be regarded as competing with plants for the chemical energy stored as a consequence of photosynthesis. Many plants respond by developing structures or chemical substances which make them less palatable. Egg-laying females of many insects recognize the correct larval food-plant by its distinctive aroma, which may originally have been evolved as a defence mechanism. The long-term evolutionary outcome is that each plant species is exploited by a relatively small number of species of herbivores, and each species of herbivore exploits a relatively narrow range of similar plants. There is in addition evidence that even within a species of plant there are individual genetic differences which determine palatability. Experiments performed in the United States show that when scale insects, *Nuculaspis californica*, feeding on ponderosa pine, are transplanted from one tree to another, their survival rate is low and they leave few offspring; but the survival rate of transplants elsewhere on the same tree is unaltered. This suggests that these highly sedentary sap-sucking insects are adapted to individual trees and not simply to the species as a whole. In Nigeria the leaves of different cultivated varieties of cassava (a common tropical crop) vary in their resistance to attack by the grasshopper, *Zonocerus variegatus*; indeed the varieties of many species of cultivated plants differ in their ability to resist insect attack.

The peculiarly restricted distribution in Britain of the black hairstreak butterfly described in Chapter 1 can now be re-examined in the light of the above remarks. It will be recalled that blackthorn, the larval food-plant, occurs throughout Britain, but that the butterfly occurs only in a small area. Is it possible that the tissues of blackthorn leaves differ in chemical composition in different parts of its distribution and that the black hairstreak is adapted to one particular chemical variety? This is certainly a feasible explanation for the strange distribution of the butterfly. The restriction of the black hairstreak to only part of the range of the blackthorn may represent just one step in the continuous evolutionary jostling between eater and eaten. If this interpretation is correct, the outcome might be either a spread in unpalatability of the blackthorn so that black hairstreaks become extinct, or an extension of the butterfly's distribution as it adapts to eating other chemical strains of blackthorn. We know that the black hairstreak has been confined to its present range for at least a hundred years, which means that it may take hundreds of years before there is a significant change in its distribution.

Some herbivores are able to exploit a wide range of plant species. The caterpillars of several common moths in English gardens appear to be able to eat the leaves of almost any species of plant, native or introduced. Such caterpillars must be adapted in some way to accommodate the chemical differences in plants. The spittlebug, *Philaenus spumarius*, has been recorded as feeding on an extraordinary number and variety of plant species. In a garden at Leicester it has been found feeding on 96 of the 289 species of plant present, but has not been found on grasses, 24 species of which occur in the garden. Immature spittlebugs extract water and nutrients from the xylem (the water-conducting vessel of plants), which is unlikely to contain unpalatable chemical compounds, and hence by exploiting what seems an unpromising source of food the bugs are able to use a wide range of different species of plants. The use of many different plants by a herbivore is just as much an adaptation as is restriction to a narrow range of species.

### Buddleia in Europe: adaptation or adjustment?

European gardens contain many plants introduced from distant parts of the world, but only exceptionally do such plants escape and become common in the wild state. One that has done so is buddleia, *Buddleia davidii*, a member of the Buddleiaceae, a family unrepresented in the native flora of Britain and Europe. In about ninety years the plant has

*Plate 19:* A *Buddleia* thicket colonizing a building site.

become naturalized and common: conditions in Europe are evidently favourable for buddleia, and its present abundance introduces a totally new element into the array of potential food-plants available to herbivores.

Buddleia comes from the highland areas of west and central China. Père David, a missionary and naturalist, sent a sample to France in 1869, and seed sent from France to Kew Gardens in London in 1896 produced the first plants of a species that was later to colonize almost the whole country, although it is probable that today's buddleias derive mainly from later introductions in the first decade of the present century. The newly-introduced shrub attracted immediate attention. It was lauded in the gardening magazines of the period for its scented white, pink, or purple flowers, and a host of varieties differing in growth form and flower colour became available. The recent surge of interest in wildlife in gardens has drawn attention to buddleia or 'butterfly bush' whose flowers, even in a suburban garden, attract up to twenty species of butterfly as well as many species of bees, hoverflies, and other insects. Most of the varieties are hardy in Britain and Europe and retain some leaves in winter, an added attraction for a town garden.

The fascination of buddleia is the speed and extent of its assimilation into the native flora. In Britain it started to spread in the 1930s, and from the late 1940s onwards became a familiar sight on waste ground and bombed sites. Today it immediately colonizes abandoned building sites and derelict land where there is plenty of lime in the soil. It can form tall, dense thickets, calling to mind a nineteenth-century description of its native growth in China, where it was said to provide good cover for leopards.

Why has buddleia spread so rapidly and effectively? The leaves have a fine downy covering which probably stabilizes the micro-environment around the pores in the leaf surface, across which gas exchange takes place; this may facilitate photosynthesis and respiration in polluted environments. Although the plant does not require high concentrations of lime in the soil, it is extremely lime-tolerant, enabling it to exploit the vacant niches offered by builders' rubble which other species cannot exploit. For a woody plant it grows rapidly and can reach a height of over a metre in its first year, which ought to make it a good competitor with the few other species that can colonize recently derelict land. It seems, then, that there was a vacant niche into which a large, woody, lime-tolerant shrub could fit.

After only ninety years in Europe we might not expect to find herbivores feeding on such an exotic plant as buddleia. Leaving aside the many nectar feeders attracted to its strongly scented flowers, the list of primary consumers is impressive and includes the caterpillars of more than a dozen species of moths, and, in both Britain and Holland, the caterpillars of the holly blue butterfly. All of these moths, and to some extent the holly blue, feed on a wide range of plants, and have presumably added just one more species to their diet. In recent years there have been many reports of the caterpillars of the mullein moth, *Cucullia verbasci*, feeding on buddleia leaves. The caterpillars of this species are normally restricted to mullein and figwort and their occurrence on buddleia immediately suggests a chemical similarity between the Scrophulariaceae (to which mullein and figwort belong) and the Buddleiaceae. Indeed it is possible that the moths have detected an evolutionary relationship between the two plant families which thus far has eluded botanists.

We do not know if the mullein moth has changed and become adapted to feeding on buddleia, or whether because of the presence of similar chemical compounds in the two plant families there has simply been an opportunist adjustment. Nor do we know if buddleia itself has changed genetically since it escaped and colonized the vacant niches on building sites and similar places. As already mentioned there are numerous (genetic) strains of buddleia, and it is probable that some survive better in a wild or semi-wild state than others. There is considerable scope for a thorough survey of buddleia in order to find out which strains flourish best in different environments and what kinds of herbivores are able to feed on its leaves, flowers, and seeds.

## Some genetic consequences of pollution

Pollution constitutes an environmental change and industrial man is therefore a potential instigator of evolution. The environmental effects of industry are numerous and we shall here confine our attention to how industrialization can alter species. The phenomenon to be described is known as industrial melanism and it involves the replacement of pale-coloured moths by dark-coloured ones in and around centres of heavy industry. The way in which this works is of great theoretical interest and is perhaps an illustration of how, instead of disappearing because of the alteration of the natural environment, species are able to adapt themselves to the changed circumstances brought about by man, through the process of natural selection.

Industry produces atmospheric pollution and among the first organisms to suffer the effects of this are lichens, mosses, and other small plants growing on the trunks and branches of trees. These plants give a tree trunk its distinctive colour and pattern, and if they disappear the bark of the tree is dark brown, almost black. When there is very heavy pollution there may also be a deposition of soot on tree trunks which results in further blackening. If you live in an industrial city, try rubbing your hands on the trunk of a tree and you will find that there is plenty of accumulated soot. Tree trunks provide day-time resting places for many species of night-flying moths. The wings of these moths are normally intricately coloured and patterned and match the lichens and other plants that usually grow on tree trunks. Birds search for and eat any moths they find, but the coloration of the moths makes them difficult to detect because it blends perfectly with the background on which they rest during the day. But when the tree trunks have lost their covering of lichens the advantage in the normal coloration of the moths disappears and they become conspicuous and easy to find.

In the early 1900s several species of North American moths suddenly started producing black, or melanic, forms that lacked the usual intricate pattern. These melanics first appeared in and around centres of heavy industry such as Pittsburgh, New York, and Detroit. They are genetically determined and it is possible they have been produced for thousands of generations in the past as disadvantageous mutations. But with the disappearance of the tree-trunk flora melanic individuals are at an advantage because they match well their polluted surroundings. Industrial melanism has now affected dozens of species of North American moths and nowadays more than ninety per cent of the individuals of some species living in large cities are melanic. The melanics spread rapidly in and around polluted environments but not in unpolluted areas where the tree trunks have retained their lichen flora. A parallel change has occurred among similar western European species, starting in industrial England about fifty years earlier than in North America, which is expected as industry developed earlier in England. Pollution itself does not cause the production of melanics but through natural selection imposed by insectivorous birds it has favoured the spread of melanics in polluted areas. Birds may not be

*Plate 20:* Industrial melanism in moths. The left hand picture shows pale and melanic moths on the 'original' light background of a lichen-covered tree trunk. The pale moth, just below and to the right of the melanic form, is very difficult to see. The right hand picture shows the same two forms on a polluted tree trunk near Birmingham, England. This time it is the melanic form, to the lower left, that fades into the background.

entirely responsible for the increase in melanism as some melanics appear to be hardier than pale individuals, possibly because their caterpillars are better able to cope with contaminated food-plants.

Other insects besides moths have been affected, among them the two-spot ladybird. Some populations of these ladybirds now have high frequencies of melanics in and around industrial cities in England. But it is less likely that natural selection by birds is important as ladybirds are conspicuous insects and unpalatable to predators. Recently the level of pollution in some areas of England has been reduced. Already the moths and ladybirds are responding, and there have been slight decreases in the frequency of melanics. It therefore seems that melanic moths and ladybirds are sensitive indicators of the level of pollution.

## Natural selection and stability

At the beginning of this chapter we reasoned that natural selection must be occurring all the time but that only under rather exceptional circumstances will it lead to evolutionary change and adaptation. All the examples we have discussed have been initiated by man's alteration of the natural environment or by deliberate efforts at cultivating and domesticating wild species of plants and animals. It is indeed difficult to find direct evidence of evolution by natural selection where man is not implicated, but this does not of course mean that man is necessary to the selective process; rather it suggests that natural modifications of the environment occur slowly, so slowly that we cannot expect to find evidence of evolution in most situations. The fossil record, which is remarkably detailed, shows that although some groups of organisms have evolved relatively quickly, others have remained unchanged for hundreds of millions of years. Evolution may occur quickly or slowly, it depends on the circumstances and the organisms involved, and there are no rules which determine its rate. Natural selection occurs all the time and its effects are mostly to keep organisms just as they are generation after generation.

In Chapter 1 we described the consequences of mowing a grass lawn. Among other things the lawn mower chops flower heads off grasses and weeds and consequently prevents them from producing seeds. But flowering may occur on stems so short that they miss the mower and in this way some seeds may be produced. If the capacity to produce flowers on short stems is inherited (which it may be) the lawn mower is acting selectively and favouring those plants. Gardeners will agree that weeds growing in lawns tend to produce flowers close to the ground compared with the same species of weeds growing elsewhere and the possibility of a lawn mower exerting selection is therefore considerable. This example will serve to introduce the notion that natural selection can impart stability to a population in terms of the population's genetic variation, but we can consider some other examples.

Parents are usually pleased if their baby is born at a weight near the average; unusually light and very heavy babies are regarded as less desirable, and for good reason, because those at both extremes have less chance of survival than babies born at the average weight. As with all aspects of natural selection we are speaking in statistical terms and although the probability of death of babies well below or above average weight is greater than that for babies of average weight it becomes serious only in exceptional circumstances. There is however a constant small selection against babies of abnormal weight which, generation after generation, results in most babies being born close to the average weight. The parallel between babies and weeds in the lawn should be evident.

Malaria is one of the most dreaded diseases afflicting man. It is transmitted from person to person by mosquitoes as they bite and suck blood. The salivary glands of certain species of mosquitoes contain an infectious micro-organism called *Plasmodium* which is introduced into the blood stream when the mosquito bites. Only a few of the hundreds of species of mosquito transmit malaria and nowadays the disease is more or less confined to tropical countries. Unlike many human diseases, particularly those caused by viruses, malaria does not appear and disappear and there are rarely epidemics. It tends to persist for years at roughly the same intensity and only exceptionally are there major outbreaks, and although of course it can spread into new areas it does so slowly and cannot be compared with diseases like cholera and influenza which spread rapidly.

*Plasmodium*, the organism causing malaria, is a parasite that lives and undergoes phases of its life cycle inside the red blood cells. Malaria frequently causes death, especially among children, and as would be expected from what we know about natural selection, the probability of death is not the same for all individuals who are suffering from the disease. Some people acquire a natural immunity after repeated exposure; others are genetically more resistant, and this is what interests us here.

Haemoglobin is the respiratory pigment that gives blood its distinctive red colour. A human embryo contains a type of haemoglobin known as haemoglobin F and this is replaced by the normal type A soon after birth. In some people another type, haemoglobin S, partly or completely replaces A. The partial or almost complete substitution of A by S is an inherited condition and individuals with an almost complete replacement usually die in childhood of anaemia because the red blood cells collapse and tend to block the capillaries. Individuals with some A and some S are normal, unless they are exposed to reduced oxygen pressures (unlikely for most people), but they are more resistant to malaria than those with A alone. Thus in areas where malaria is common people with some haemoglobin S survive better than those with 'normal' haemoglobin A, and of course much better than those with mainly haemoglobin S. This leads to an equilibrium in the frequency of various kinds of people that can be interpreted in terms of natural selection. The malarial parasite is the selective force. The likelihood of infection with malaria depends on various factors, including the abundance of suitable mosquitoes (which in turn may be determined by rainfall, patterns of agriculture, and so on), the density of the human population (which affects the rate of transmission of the disease), and nowadays whether drugs are being used to combat the disease. In many parts of Africa where malaria is persistent people with the inherited immunity survive better than those without it and the frequency of such people in the population depends on the intensity of malaria in the area. In some parts of Africa up to forty per cent of the population have haemoglobin S but in other areas it is rare or absent, its precise frequency being determined by selection imparted by malaria. There are other kinds of abnormal haemoglobins and it is believed that these are also associated with malaria, although not necessarily with the same kind of malaria correlated with the presence of haemoglobin S.

## Dutch elm disease

The English countryside has changed drastically in appearance over the past decade as elm trees have died and have fallen over or been felled. The characteristic tree of hedgerows and shelter belts around farmhouses for hundreds of years was the elm, usually the English elm, *Ulmus procera*, or the wych elm, *U. glabra*, and occasionally other species or hybrids. Of an original total of about thirty million trees, eleven million had been lost by 1978 as a result of Dutch elm disease which still shows no sign of abating. The disease is caused by a yeast-like fungus carried in the sap of the tree. A toxin produced by the fungus causes the walls of the sap-carrying vessels to develop enlargements which block the flow of nutrients. As the fungus spreads through a tree, leaves yellow and wither, more and more branches die, and eventually the tree succumbs. The fungus is transported from tree to tree by small beetles of the genus *Scolytus*. Newly-emerged beetles feed on elm sap, and after mating the females burrow and excavate a tunnel between bark and wood, laying eggs alternately on each side. The larvae eat their way through the bark at right angles to the original tunnel, producing a characteristic radiating pattern of feeding galleries which are clearly visible when bark is stripped from infested trees.

The fungus is probably of Asian origin, and the disease-causing strain was discovered in France in 1818 and first noticed in Britain in

1927. The disease spread but apparently in a mild form, and although perhaps twenty per cent of English elms were killed the epidemic began to decline in 1937. However, a more harmful strain appeared in North America in the 1960s and reached Britain before 1970 apparently with imported timber. All species, varieties, and hybrids of elms are affected, but the wych elm (the only native elm) is more resistant to the disease and there is no question that the English elm is most susceptible.

The surest way of stemming the spread of the disease and saving uninfected elms from death is the total destruction by felling and burning of diseased trees; but this is expensive and infection is not always evident early enough for action to be taken. There have been attempts to control the disease either by injection of the trees with fungicide or spraying with insecticide to deter the beetles, but neither method has had much success; legislation to stop the movement of timber around the country goes some way to control the spread of the disease but is difficult to implement. The virtual eradication of the English elm now seems inevitable.

The best hope for the future appearance of elms in the countryside lies in planting trees resistant to fungus infection. As described in Chapter 3, the English elm rarely or never sets fertile seed and habitually reproduces by vegetative means. Consequently most or all of the elms in an area, certainly in a single hedgerow, are genetically identical and often retain physical continuity through their root systems; furthermore it is not unusual for the roots of different trees to graft together where they touch. This has two consequences for the spread of the Dutch elm disease: the fungus can spread through the root system from one tree to the next, and the chances of production of a genetic strain of English elm resistant to the disease are remote. Were English elms genetically distinct individuals reproducing by a sexual process it would be possible through natural selection for a mutant strain resistant to the disease to increase in frequency at the expense of the non-resistant strain. In other words the English elm could adapt to the fungus. The wych elm regularly produces fertile seed and does not reproduce by root suckers as frequently as the English elm. Unlike English elms, wych elms are usually true genetic individuals, and in the long run they stand a better chance of survival. The susceptibility of the English elm to Dutch elm disease is in part a consequence of its means of reproduction and the associated inability to adapt to the new features of the environment created by the fungus.

## A man-made world

With the continual introduction of new methods of agriculture, and new strains of crops and domestic animals, and with the widespread destruction of natural ecosystems and the growing threat of industrial pollution, we can expect to discover more and more examples of man inducing evolutionary and adaptive changes in himself and other organisms. The most striking examples of selection by man to produce organisms adapted to a man-made world are the strains of wheat, rice, and other cereals that constitute the so-called 'green revolution' crops. These crops give a high yield, concentrating most of the input of energy and nutrients into grain and using the minimum for vegetative growth and maintenance. But high yields can only be sustained when there is a massive input of fertilizers, pesticides, and weed-killers. A large energy subsidy of this sort is expensive and beyond the means of many poor countries for whom the green revolution held promise of increased food supplies. The green revolution has produced crop varieties adapted to the intensive, energy-subsidized farming of the affluent, but which seem of dubious value for peasant cultivators who have little or no access to energy subsidies.

The flowering, and hence fruiting, of certain temperate plants can be brought forward by subjecting the germinating seeds to cold treatment, a phenomenon known as vernalization. Crop plants can be induced to develop much more quickly by such means; winter wheat, normally sown in the autumn, can be sown in spring and complete its development in one growing season if vernalized before planting. The process has been extensively used in the U.S.S.R. to avoid damage to wheat seeds and seedlings in severe winters and to extend the northern limits of the range of grain crops. The Soviet geneticist, Lysenko, thought that vernalization caused adaptation; he revived the beliefs of the nineteenth-century French biologist and philosopher, Lamarck, who maintained that characteristics acquired during an individual's lifetime can be inherited. Few people nowadays accept this, but Lysenko based his programme of vernalization of wheat on Lamarck's views on the inheritance of acquired characters. Disastrous crop failures in the U.S.S.R. were blamed (rather unfairly) on Lysenko's genetic theories, and in the end he was politically and scientifically discredited. The essence of Lysenko's personal failure was lack of distinction between adjustment and adaptation. Cold-resistant wheat strains could be selected to produce an adaptation to a short growing season, but so far as is known a plant's adjustment as a consequence of vernalization cannot be passed to its offspring: the process must be repeated with each new generation of seed.

If you have a garden and particularly one in a suburban environment, your view of the living world will be dominated by the products of man's inventiveness. The holly and the privet bushes in the garden are probably not exactly like those that grow in the wild, most of the ornamental flowers are unlike their nearest wild relatives, and the vegetables you grow are not plants you would expect to see occurring naturally. One effect of man-induced modification of the environment is the increasingly rapid disappearance of wild species of plants and animals. It has been predicted that by the end of the century thousands of species will have gone; in tropical forests alone one species is reckoned to disappear every day. Such species have no place in a man-made world and rarely have the opportunity to adapt to it. Thus not only does man change the species-composition of virtually all communities, but also the genetic structure of innumerable plants and animals. His actions are often deliberate, but perhaps even more often are accidental, resulting from his carelessness, and desire to control and exploit nature.

Increased awareness of changes and adaptations in plants and animals has brought with it an inquisitive approach to human genetics, not only a wish to understand which traits are inherited and how, but also to understand behavioural as well as physical differences between human populations. The controversy over the relative influences of inheritance and environment on intelligence and learning ability has become a sociological and even a political issue, but it would be tragic if policies and programmes in education and community affairs were based on the same kind of ignorance as has often underlain man's attempts to manipulate the natural world.

# 7 How does ecology affect us?

Every year we hear of discoveries of fossils which shed more light on man's evolutionary history. But these fossils tell little of man's cultural and social development, and not a great deal about man's ecological relationships with other organisms. It seems certain that no matter what mental qualities early man may have possessed, he survived as a species without conspicuously interfering with the natural ecosystems of the world. Man's discovery that fire could be used advantageously was probably the first break-through to open up a whole range of new opportunities. Fire was used not only for cooking, keeping warm, and warding off dangerous animals, but for destroying vegetation, thus for the first time modifying natural ecosystems on a grand scale. About the same time man started to make and use tools and weapons further increasing his potential as an exploiter. But it was the invention of agriculture and the domestication of animals that provided the first really big stimulus for an increase in human numbers and for the alteration of natural ecosystems. At about the same time it was discovered that extracts from plants and animals could be used for medicinal purposes. Over the centuries this discovery has developed into modern medicine as we know it, and even today a large proportion of the drugs used in fighting disease are derived from other organisms. Preventive medicine came much later, and only in recent times has it had a major impact on the rate of human population growth.

It probably took hundreds of thousands of years for the human population to reach its first 1000 million, an event which occurred about 1830. It rose to 2000 million by 1930, 3000 million by 1960, and in 1976 it passed the 4000 million mark. This rapid rate of growth in recent times can be attributed to the modernization of agriculture and the discovery of really effective means of preventing and curing killer diseases. In terms of evolutionary history, man has been on earth for only a short time; there is no particular reason why he should go on for ever, or indeed for much longer. Other species have flourished in the past and have died out, and there is as yet little reason to suppose that man will be an exception. It is difficult to visualize what the world will be like at the turn of the present century, a mere twenty years away, and absolutely impossible to imagine the world in a hundred or a thousand years from now, and yet these spans of time are only a tiny fraction of the history of man, let alone the history of life.

Human populations continue to be exposed to natural selection, as we saw in the last chapter when we discussed malaria and the birth-weights of babies, but some of the selective pressures exerted now are probably different from those in the past. People who a hundred years ago would have died from genetic abnormalities before they had a chance to have children, now survive and breed.

But the most pervasive ecological feature of man is his inequality as a producer, consumer, and decomposer of the world's natural resources. Like other animals, man is essentially dependent on plants as a source of food, and there seems no reason to doubt that ultimately his numbers will be limited by the carrying capacity of the world defined in terms of the availability of food. But there are enormous individual and regional differences in man's ability and opportunity to exploit resources. Two-thirds of the world's population live in poverty while the remaining third in the economically developed countries live in comparative luxury. The poor people consume a relatively small amount of the world's non-renewable resources; many are sick and suffer from a shortage of calories and protein; and their rate of population growth is more than twice that of the rich people in developed countries. Moreover the poor are still unwilling or unable to limit their birth rate, the only possible way of initiating an improvement in their standard of living. Underdeveloped countries, especially those in the tropics, produce raw materials for consumption in developed countries and because of historical, economic, and political circumstances it seems that the vast majority of individuals in these countries will never have the chance of enjoying the fruits of industrialization.

Man's ecological inequalities come sharply into focus when we consider that a person in a rich country today can sit in front of a colour television set and watch other people starving to death. It is true that those starving will almost certainly be in a remote and unfamiliar country, but our efficient communication systems, making use of satellites orbiting the earth, provide the possibility to see 'live' other people suffering and dying simply because our political and economic systems do not permit a fair distribution of wealth, resources, and opportunities. Europeans and North Americans have been brought up in a culture which believes that the resources of the world are there to exploit. To make a profit at the expense of others is considered honourable and worthy, and the bigger the individual profit the more others are likely to congratulate a man for having 'got on' in life. Economic growth is sought after by politicians of all persuasions and in all countries, but only a few people in a few countries are sharing the profits resulting from growth. Notwithstanding political dogma, economic growth, like all exponential processes, cannot be sustained indefinitely. Many attempts at improving the human situation are short-sighted and inward-looking. It often seems that nations care little about each other, except as sources of raw materials and places where goods can be bought and sold, and we are now experiencing the biggest economic scramble the world has ever seen.

We can return to specific issues later, but before doing so let us ask ourselves whether a knowledge of ecology can help in understanding the pressing problems of the world as they now appear and whether ecological thinking can offer hope.

## Thinking ecologically

The first important lesson to learn is that man is part of nature and that the rest of nature was not put there for man to exploit, the claims of business, political, and religious leaders notwithstanding. In 1894 T. H. Huxley wrote an essay called 'Man's place in nature', and others have since argued that as we are a product of evolution it is legitimate to consider ourselves in the context of the rest of the living world, no matter what special properties, spiritual or otherwise, we may attribute to ourselves. If, then, we acknowledge our evolutionary origin, we can rather more easily try and put ourselves into ecological perspective; and in attempting this we shall draw on the principles discussed in the preceding chapters.

It is an axiom of ecological theory that all organisms modify to some extent the ecosystems in which they live. Organisms are of course part of ecosystems, and the presence of this or that individual or species is bound to affect the way in which an ecosystem works. The spectrum of species in an ecosystem depends on a variety of factors, including climate, the amount of light, and the availability of inorganic materials (in both soil and water) which determine the productivity and diversity of photosynthetic plants. The plants and the animals that feed on them are capable of rapid population growth, but this tends to occur only when there is a breakdown of the natural balance of the ecosystem. It is, as we have seen, more likely to occur in 'simple' ecosystems, such as those at high latitudes or in environments unfavourable to all but a few species. Evidently complexity (in terms of species diversity) is correlated with stability. All tendencies for growth in numbers or consumption are necessarily temporary and sooner or later they are halted and stabilized by pressures exerted by the environment.

There is no reason to suppose that man's present rapid rates of population growth and consumption will remain immune from regulation, and unless there is a man-induced catastrophe like a nuclear war there is every reason to suppose that stability will be brought about in a density-dependent manner in the not too distant future. The question, of course, is how exactly this will occur, and when. It is particularly clear that our present consumption of oil will have to come to an end soon, as at the present rate there will be none left worth exploiting in the very near future. The price of petrol may soon become so prohibitive that the ordinary motorist will be unable to afford to run a car, although it must be admitted that the substantial increases in price since 1973 have not resulted in less being used. Man's use of metals, at present an extremely wasteful process, could be reorganized in such a way that a substantial amount of material now thrown away is recycled in much the same way as materials are cycled in natural ecosystems.

Attempts to increase agricultural productivity to provide food for an expanding population merely postpone what is inevitable, as there is ultimately a limit to what can be produced that is suitable for human consumption. Indeed one could argue that the more food produced the worse the long-term prospects for mankind: the bigger the population the more dramatic its crash. We should therefore be on the look-out for events likely to control human population growth.

There are several possibilities. One is that man deliberately limits population growth by controlling the birth rate, approaching the problem in much the same way as he has in controlling the death rate. This could be (and is) successful in developed countries where there is an individual stimulus because of the possibility of improving standards of living by having smaller families, but for a variety of social and religious reasons (and we should include ignorance) birth-control as a means of limiting population growth is a non-starter in the poorer countries, although China is possibly an exception. Apart from any other considerations, birth control is seen as something of a white man's plot by many of the more influential leaders in the under-developed world. No one can really expect birth-control to have a significant effect on the rate of world population growth when the majority of the people in the world are unaware that there is a problem or that to control births is possible. In some countries with high rates of population growth more than ninety per cent of the people are illiterate, a situation which is not promising if we want to introduce such a delicate and intricate idea as birth-control.

It is far more likely that the growth in the human population will be stabilized through the operation of density-dependent factors. A shortage of food, and perhaps disease, is likely to create a situation where deaths begin to exceed births, although it is a matter of conjecture when and where this will begin on a large scale. Nearly every year we hear of famine in India and severe food shortage in other parts of the world. Drought is frequently blamed for the failure of crops, but drought itself is often induced by man's destruction of the natural vegetation in his attempts to cultivate and to graze animals. Already vast numbers of young people die of disease, and the effects of both disease and famine are further aggravated by wars between nations and within nations which necessarily result in a breakdown of health services and agricultural productivity. The deliberate destruction of rice fields in the Vietnam War is but one example of man's nastiness to man.

A misleading air of confidence is maintained by health authorities, including the World Health Organization, about the extent to which killing diseases are under control. Outbreaks of such diseases occur repeatedly. We hear about them when they affect a developed country or a country in which we have special interests, but we have yet to witness the more dramatic possibilities. The black death or plague of the Middle Ages and the influenza epidemic of the First World War are popularly believed to be events of the past which modern medicine will prevent from occurring again. But it is probably only a matter of time before yellow fever spreads from Africa to the teeming millions in

India. Yellow fever, which is caused by a mosquito-borne virus, erupts periodically in Africa, where there is evidence of a spread to the East Coast. Once it arrives there, it may have little difficulty in crossing to India, where it would have a terrifying impact on the non-immune population already suffering from food and protein shortage and numerous diseases. Yellow fever (like all virus diseases) has more disastrous consequences in densely crowded populations than in areas where people are thin on the ground.

The foregoing remarks are simply descriptions of what is happening and of what is likely to happen. By thinking ecologically we have the advantage of being able to anticipate some events, but we have not allowed for the more catastrophic possibilities of wars fought with nuclear weapons or the deliberate introduction of killing micro-organisms or toxins into an unprotected population. These possibilities are real, and as the scramble for what is left of the world's resources and space intensifies we should not dismiss them as unlikely. If and when they do occur, the resulting mortality is likely to be more density-independent than the 'natural' events just discussed.

*Plate 21:* Open-cast mining. Its effects are extremely destructive to natural ecosystems, as can be seen in the photograph. Metals and other materials are redistributed around the world and are effectively 'lost'.

Man is the only organism to consume fossil fuels as a source of energy and substantially to re-distribute metals and other materials around the earth. The process by which fossil fuels and metal ores are obtained and transported tends to have a destructive effect on natural ecosystems and on other species of plants and animals. The wastes generated by industry result in environmental pollution affecting ourselves as well as other organisms. But man's most striking contribution towards an imbalance of nature arises from cultivating plants and domesticating animals. Food is produced not only to feed people but for profit. Despite disclaimers, the possibility of profit is the main stimulus for improving farming methods by the creation of monocultures of selected and desirable species of crops and animals which can be efficiently harvested, packaged, and sold. Indeed farming in some regions is now so efficient that huge surpluses accumulate which, however, do not lead to a fall in price to the consumer. When the first edition of this book was being written, there was a scandal developing over the vast stocks and high prices of butter within the European Common Market which seemed unlikely to be resolved in favour of the consumer; the outcome was that prices for farmers were guaranteed, butter prices remained high within the Common Market, and a substantial quantity was sold cheaply to the U.S.S.R., which promptly re-sold some of it at a profit to Common Market countries. By 1978 the supply of butter and other milk products was so much in excess of demand that even farmers were amenable to consideration of the removal of guaranteed prices.

All over the world traditional methods of agriculture — growing crops for oneself — have been, or are being, abandoned, and the land developed for the large-scale production of commercially important crops. Underdeveloped countries are being pressured by developed countries to produce more 'cash crops', to the detriment of products for local consumption and to the quality of the land. In ecological terms the creation of monocultures means that the natural diversity of ecosystems is reduced by eliminating all but the desired species. As we have shown earlier, the number of species in an ecosystem is determined by the nature of the environment. When an environment is simplified by the development of a monoculture most species disappear as their resources diminish, but a few are able to exploit the new circumstances. This is why we recognize some plants as weeds and some animals as pests. Monocultures are potentially less stable than natural ecosystems because they contain far fewer and less variety of species. To convert an area of rain forest to a rice field is to replace an organized and stable ecosystem that has taken thousands of years to put together with a simple monoculture dominated by one species and the organisms that happen to be able to exploit that species. The result is that man is constantly faced with the depredations of weeds and pests able to adjust themselves to the monoculture he has created. This in turn leads to campaigns designed to destroy the weeds and pests. There are many ways of doing this, the most efficient and cheapest (and therefore the one likely to yield the biggest profits from the crop) being to poison the environment just enough to destroy noxious species and to leave the crop unaffected, or at least superficially unaffected. Hopefully poisoning will not affect potential consumers of the crop, but this is not a major issue with the grower, and the fact that it does so

*Plate 22:* A ship loaded with toxic chemical wastes intended for dumping in the North Sea is prevented from leaving Esbjerg harbour (Denmark) by fishing boats. The fishermen were concerned that dumping chemical wastes would endanger their fishing interests, but after it was agreed that the wastes would be dumped in the Atlantic the fishermen allowed the ship to leave. Thus what seemed to be an effective form of protest turned out to be nothing more than people protecting their own environment, and the fishermen did not seem to care about what happened in the Atlantic.

from time to time should be no cause for astonishment. Poisoning the environment with pesticides also has unexpected ecological consequences, among them the disappearance of organisms not themselves pests, often in areas far removed from the scene of operations. On the other hand, from what we now know of the intricacies of food webs and the ways in which materials are recycled in nature, we can be sure that poisoning an environment has widespread effects elsewhere and, it might be added, often leads to the evolution of pesticide resistance in the organisms it is designed to eliminate.

We are speaking very generally about the living world, and we have merely attempted to place man's activities in an ecological framework which can be understood. The lessons of density-dependence seem obvious enough, as do the limits to population growth and consumption of resources. The replacement of complex ecosystems by simplified monocultures and the poisoning of the environment considered necessary to maintain these monocultures has resulted in a large-scale re-assembly of the living world. Yet all of this has happened in virtually no time. Only in the past 10,000 years has man had a significant effect on the world, which is only 0·5 per cent of the time that man as a species has been around. Only in the last 100 years has the impact of man on the natural world begun to be appreciated and it is only in the last twenty years that we have really started to question the wisdom of what we are doing. It is believed that there has been life on earth for at least 3000 million years; man has been on the scene for only 1/1,500 of this enormous time-span and his disrupting activities have been going on for only 1/300,000 of this period, if we accept that disruption began 10,000 years ago with the beginnings of agriculture. The ecosystems of hundreds of millions of years ago were not, of course, the same as now, but their transition has been gradual, taking millions of years, while man's impact on the global environment has occurred in what, in evolutionary terms, is no more than a few seconds.

Of course we are all encouraged to live in the hope that science, technology, and business skills will overcome all our difficulties, just as, it is claimed, similar difficulties were overcome in the past. But no amount of science and technology will automatically lead to a stable population and to zero economic growth, and it looks as if to a large extent we shall have to accept what is happening. Even if a few developed countries manage to reach population and economic stability, the rest of the world would not follow suit, and any advantage gained would quickly be lost. We have already mentioned man's ecological inequality, a problem which has political, moral, and economic aspects as well, and which is likely to have increasing repercussions on those of us that enjoy the luxuries of industrialization. Political leaders in poor countries are fast becoming aware of the extent to which their countries have been, and continue to be, exploited by the business interests of the industrial nations, and they can be expected to tighten the weak hold they at present have on the natural resources of their countries. As this happens, the rich nations will have to pay more for raw materials. Economic and business associations like the Common Market will become more powerful as the scramble for favours and opportunities for access to raw materials intensifies. Already industrial man's acquisitiveness has led to suspicion of the motives of others, not only of people of different nationalities and ethnic groups, but also of neighbours and associates, and certainly of those few who question the wisdom of continued economic expansion.

## Man's ecological inequality

Whether we take the view that the present rate of growth of the world's population and the rate of use of natural resources will bring disaster by the turn of the century, or that somehow modern technology will solve our problems, it is evident that the present affluence of the industrial nations, with about a third of the world's population, can be sustained only if the remainder of the world continues in the role of the producer of raw materials. These raw materials include crop products and minerals, but especially oil. Many, but not all, of the producer countries are officially designated (by the United Nations) as underdeveloped (the so-called Third World) and are characterized by poor or non-existent industries, high rates of population increase, illiteracy, ill-health, food shortage or nutrient deficiency, and generally low standards of living, at least when compared with the consumer countries. In 1977 the World Bank (which keeps information on such matters because it hands out money to the governments of poor countries) estimated that 1200 million people still lack access to safe drinking water or health care, 700 million do not have enough food, and 250 million living in urban areas are inadequately housed. These figures may not be entirely accurate (other organizations produce different figures) but they at least reflect the magnitude of world poverty. Most underdeveloped countries are in the tropics and were until recently colonies of European powers. They now receive some aid and much investment from their former colonial masters and from other rich countries who have realized their potential for economic exploitation. Aid and investment are claimed to promote economic development, but nearly always they involve nothing but increased exploitation of natural resources. If you travel to and from Africa you may meet well-dressed European businessmen who after the usual exchange of pleasantries will make remarks like 'Business is good in Nigeria' or 'We are expanding our interests in Kenya'. Thus it may be that most African countries are now enjoying a shaky political independence, but there is no question of economic independence.

The Europeans brought to their colonies literacy, the rule of law, firearms, Christianity, methods of curing and preventing killing diseases, but above all the notion that the resources of the environment should be exploited for profit. By a process of trial and error and in environments that were poorly understood, monocultures of cash crops were established, the products of which could be sent to Europe with the minimum of preparation and processing. Labour was and still is cheap and plentiful, and the methods of cultivation inefficient, but huge profits were made by enterprising men who knew the expanding markets of Europe. Minerals were located, and as the demands for them in the growing industrial complexes of Europe increased the scramble was accelerated. Needless to say the industrial countries, now so heavily dependent on other people's resources, are becoming uneasy and suspicious of each other's motives as competition for the rights to exploit the resources of underdeveloped countries intensifies.

The Europeans have also spread themselves around the world and in many areas, notably North America, South America, Australia, and New Zealand, the indigenous people have been reduced in numbers by a variety of methods, including war, confiscation of land, accidental introduction of disease, and in places like Brazil by hybridization on such a scale that local identity has been almost lost. In other areas like South Africa the indigenous people have been inhibited from making use of their own resources and their own environment.

The new Brazilians (of European, African, and hybrid origin), together with international entrepreneurs from a wide array of countries, are systematically destroying the life-style of the indigenous Indians in the interests of 'development'. Their approach is to confiscate land, cut down forests, build roads, destroy villages, and to open up the country in much the same way as the Europeans did in North America in the pioneer days of the nineteenth century. Brazil enjoys a phenomenal rate of economic growth but the cost in terms of destruction of natural ecosystems and degradation of human life is enormous. A few people both in Brazil and outside believe that this form of 'cowboy' economy will disrupt the recycling of water and other essential materials to an extent that within thirty years the land will become a biological desert which will not even support the brand-new cities that have been or are being built.

Such activities as these began several hundred years ago and perhaps reached a peak in the second half of the nineteenth century. It was not until after the Second World War, when most colonies had gained independence, that it was felt that the days of exploitation were over. But this was not so. With the establishment of independent nations the new rulers found themselves controlling the destinies of populations which included the poorest people in the world. Each nation found itself heavily dependent on the export of one or more plant product or mineral. For some commodities, including cocoa and coffee, marketing boards were set up to determine and restrict production and export. Thus even if a country is capable of producing more it is not allowed to do so: some African countries have even been 'fined' for producing too much coffee for the world market.

A curious agglomeration of African, Caribbean, and Pacific (A.C.P.) countries have a five-year economic agreement with the European Common Market, called the Lomé Convention, which expires in 1980. The idea is (or was) for the A.C.P. countries to benefit from the expanding markets of Europe, but thus far the arrangement has failed, and the A.C.P. countries have not become industrialized, but remain as suppliers of raw materials and importers of finished products, a relationship which condemns them to bad terms of trade. Even as suppliers of raw materials, they find themselves subjected to quotas as soon as they offer products covered by the Common Agricultural Policy of the European Community: for example Botswana cannot sell its beef to Europe. Moreover, when they try to sell manufactured products like textiles to the Common Market they are warned not to concentrate exports in 'sectors which are known to be sensitive'. No wonder the A.C.P. countries want to re-negotiate the terms of the agreement, but do the European countries really have any alternative to maintaining their present role? After all, if the Common Market countries are inhibited from making large profits, they will no longer be able to help the A.C.P. countries. If poor countries rid themselves of economic domination they also risk losing aid; thus they are caught in a vicious circle which in present world economic conditions cannot easily be broken. This is tantamount to saying that the rich need the poor and the poor need the rich.

Uganda is a good example of an under-developed producer country whose economy is dependent on the export of plant products, notably coffee and cotton. It has a population of over ten million, increasing at a rate of three to four per cent per year, which is high even by world standards. About three-quarters of the people over the age of fifteen are effectively illiterate. Most are peasant cultivators who grow just enough food for their families, although some produce cash crops which they sell to local marketing boards who arrange for exports. Since independence in 1962 the country has experienced a series of political upheavals, which probably had little effect on the majority of the people, but which shifted the balance of power between various tribal groups. During colonial times (and until quite recently) most of the commercial enterprises in the country were in the hands of non-Ugandans, including Europeans, but especially Asians from India, who in effect had an almost complete stranglehold on the retail business. In the early 1970s, most of the Asians were deported and their businesses confiscated. This was done on the grounds that they had failed to integrate into the community, had sent most of their profits overseas, and were interested in Uganda only as a place where they could make money and exploit the local people. All of these accusations have some validity. Uganda's economy is now in a mess, and what few links exist with the outside world are concerned with the supply of arms required to maintain power within the country and to repel supposed threats from outside. Thus in a clumsy and often ruthless way, Uganda has managed to free itself from outside business domination, but at enormous cost.

By 1980 there will be about 458 million people on the continent of Africa, most of them very poor indeed, some starving, and many suffering from malnutrition and disease. By 1980 there will be about 496 million people in Europe, most of them relatively affluent, all of them consuming disproportionate quantities of the world's resources and dependent largely on raw materials imported from poor countries.

They will not be suffering from food shortage and they will be preoccupied with economic inflation, especially in the cost of food. Twenty years later at the turn of the century there will be 860 million people in Africa and 571 million in Europe. Africa will thus have overtaken Europe because of its much faster rate of population increase. What kind of economic situation will exist in the year 2000? No one knows, but it is worth thinking about, and if your conclusion is that the present inequalities will have been maintained, add another fifty years and ask yourself the same question, making the same assumptions about population growth and demands for resources.

Ecological inequality in the world means that Americans and Europeans (and we may add the Japanese) consume the world's resources at a rate far greater than the people in poor countries. The average European consumes steel at a rate 600 times as great as the average African, and if everyone had the opportunity of using oil at the same rate as the average American there would be no oil left in just a few years. With one or two exceptions, there is a marked correlation between countries with high birth rates and those with low per capita energy consumption. Thus the birth rate in Burundi is more than three times that in the United States while the per capita energy consumption is more than one thousand times higher in the United States than in Burundi. Such a correlation could be used to argue that a high standard of living (as indicated by a high level of energy consumption) leads to a drop in the birth rate. It is assumed that people are prepared to accept small families provided they can acquire material comforts, and it is therefore thought that the most obvious way to reduce the rate of population growth in poor countries is to improve the standard of living. This thinking is implicit in many aid projects, and forms part of the philosophy behind the export of technology and know-how from the rich to the poor. But to raise the standard of living of the world's poor to the level of the United States and Europe is impossible: the resources would be spread so thinly that everybody would be poor, a solution obviously unacceptable to the already-rich. Per capita energy consumption is but one way of judging living standards. If literacy, life expectancy, and infant mortality rates are taken into consideration some economically poor countries such as Sri Lanka and Guyana have high living standards. Evidently what is needed is a more realistic assessment of what is meant by living standards and the quality of life. You do not have to be rich to be healthy and happy.

## Producing food and preventing disease: are we succeeding?

No matter how standard of living is measured, there is no doubt about the inequality that exists between one part of the world and the other. There is no doubt, too, that a straightforward re-distribution of wealth cannot offer an acceptable or workable solution to the problem. Two aspects of life are of special significance in differentiating rich countries from poor countries: access to food and prevalence of infectious disease. In the 1960s and 1970s new discoveries in agriculture, nutrition, medicine, and pest control brought tantalizingly near the prospect of adequately feeding the people of the world and of eradicating many fatal and debilitating diseases. But hopes have recently receded: estimates of the extent of the problem differ, but it is clear that every year there are more hungry people and we are certainly not winning the battle to control infectious disease. What has gone wrong?

The financial and energy cost of green revolution crops and the cost in terms of land deterioration of ill-considered tree-felling and irrigation schemes are the main causes for reduced optimism about solving world food problems. Moreover, modern 'efficient' methods of cultivation of staple grain crops require large expanses of land and employ relatively few people; the hungry millions are often urban dwellers and do not live in the place where their food is produced — they may not even live in the same country. In financial terms small mixed farms are less efficient than vast monocultures, but they have the advantage of requiring people to live where the food is grown and also result in better use of by-products and what might otherwise be waste. There would be a global surplus and hence cheaper food if some of the more glaring anomalies of food consumption could be corrected. The rich tend to eat selfishly and wastefully. The average person in the United States makes use of more than 800 kg of grain a year, but seven-eighths is in the form of beef. In the 1960s Holland bought more skim milk to feed livestock than was given away as aid to poor countries. The ultimate anomaly is that fish meal made from third or fourth order consumers is fed to broiler poultry, normally primary consumers — work out the food chain and energy transfer involved here. Not only do some countries eat extravagantly, but they also divert potential food to other sources: Brazil aims to reduce its bill for oil imports by converting millions of tons of sugar cane and cassava to alcohol which can be used for industry and as a supplement to motor fuel.

The 1978 harvest in England and Wales surpassed previous record yields, exceeding even optimistic forecasts, but still Britain imports half of its food because of dietary habits and preferences. Food imports continue to depend upon cheap labour abroad. A tin of pineapple from Kenya sells for little more than the same weight of home-produced apples. The irony is that were the Kenyan farm labourer paid more he could afford imports, but the pineapple would be expensive and less likely to find a European market.

Whether we like it or not food production is a commercial enterprise in which the value of the yield has priority over the means used to achieve it. Surpluses, too, are commercial commodities and may be used in aid programmes without sufficient regard for the interests and health of the recipients. Thus powdered milk is frequently the mainstay of nutritional aid programmes but many adults in the tropics cannot digest lactose, a sugar present in milk, and become ill after drinking milk. Much surplus is simply wasted. It is estimated that in the United States enough food for 50 million people is thrown away each year, a large proportion of it by private households.

Food availability does not appear to operate as a factor in the regulation of the size of human populations in the same way as it operates for other animals; indeed, as we have seen, the poorest countries often have the highest rates of population increase. But we should not ignore the possibility that political and military conflict arising from dissatisfaction with the inequalities of distribution of food could in the end destroy the rich countries. The problem of hunger is readily identifiable, the capacity to deal with it exists, but what is lacking is the courage to agree that hunger is a priority problem requiring the same degree of concentrated effort that has been expended to reach the moon or to develop sophisticated weapons. The outlook is not hopeful, but a world plan to eliminate starvation and malnutrition without degrading land and wasting energy is possible. The question is whether political and economic interests will defeat the co-operation necessary to achieve an equitable system of distributing food.

The problem of preventing and eradicating disease is somewhat different. Medical aid has always been freely dispensed from rich to poor countries; indeed the export of the means of keeping people alive is far in advance of the export of the means of preventing them from being born, and is one of the most important reasons for high population growth rates in poor countries. An enormous amount of money and research has been invested in preventive medicine, and until recently it was assumed that success in controlling if not in eradicating some of the more important diseases was within reach. Thus in the early 1970s the World Health Organization waged an intensive campaign against smallpox, a virus disease similar to chickenpox but more disfiguring and often fatal. As recently as 1974 an epidemic in India caused 22,000 deaths, but by 1978 the disease was believed eradicated. A smallpox case in Somalia in 1977 was assumed to be the last, and lest a pocket of the disease remained a reward was offered for notification of any new case. In 1978 a medical school employee in Britain contracted smallpox and died; many people were quarantined and a few suffered a minor form of the disease, from which they recovered. The medical school contained one of the fourteen laboratories in the world holding stocks of the smallpox virus kept as a reference for identification of possible new infections. This was not the first time that smallpox has been caused by the careless escape of the virus from a research laboratory and we have no reason to suppose that it will be the last.

An estimated 1,000 million people live in malarial areas. In the 1950s and 1960s an all-out effort was made to eradicate malaria by spraying with insecticides to eliminate the mosquitoes that transmit the disease-causing micro-organisms, and by administering anti-malarial drugs both as a cure and as a means of prevention. In the 1950s the average annual incidence of malaria in India was 75 million cases; by 1966 the disease was well-controlled with only 40 thousand cases; but by 1976 the number had risen to over six million. In tropical Africa where the situation is particularly bleak many children contract malaria. Most recover and indeed acquire natural immunity, but when combined with other diseases and malnutrition, malaria is often fatal. Some measure of the upsurge of the disease is given by the increase in the number of cases in Britain. By 1900 endemic malaria had disappeared, although travellers from the tropics periodically brought the disease with them: in 1973 there were 336 cases, but by 1977 the figure rose to 1,527 with seven deaths. At this rate of increase we should at least by prepared for the eventual return of malaria as an endemic disease. The main reason for the resurgence of malaria is that mosquitoes and the infective micro-organisms they transmit to people are no longer so readily killed by insecticides and drugs; another reason is that people are nowadays much more mobile, which of course increases the chances of transmission.

Irrigation provides water for crops but also provides a habitat for aquatic organisms, including the fresh-water snails which are host to the larval stages of the parasite which causes schistosomiasis, a chronic debilitating disease for which treatment (when available) is unpleasant. The spread of irrigated cultivation has allowed schistosomiasis to spread and at least 200 million people are now affected. The larval stages of a biting fly, *Simulium*, which transmits small worms causing onchocerciasis (river blindness) are also aquatic. Onchocerciasis was once unknown in the Sahel region, but now a ninth of the people living in irrigated areas lose their sight as a result of the disease.

Rabies is one of the most feared diseases in the world. It is caused by a virus which invades the nervous system and produces severe inflammation. Once the infection is established it causes terrifying pain and distress, including an irrational fear of liquids, and is nearly always fatal. Foxes and other wild mammals are highly susceptible, dogs less so, although they are the usual source of infection in man. In the eighteenth century rabies was widespread in Britain; local legislation regarding dogs achieved some degree of control in the nineteenth century, although 74 people died of the disease in 1874. An Act of Parliament in 1897 gave the police power to destroy stray dogs and laws on the importation of live animals were introduced. These measures led to the disappearance of rabies from Britain in 1903, although periodically illegal imports of animals have led to short-lived outbreaks. Britain is one of the few countries free of rabies, and being an island it can remain so if legislation over imported animals is strictly enforced. But there is something of a false sense of security as in 1978 the spread of rabies across Europe brought it to within 75 km of the French Channel coast. About six million dogs live in Britain in close proximity to people; if infected animals were to be illegally shipped across the Channel the disease would soon become established, especially, it seems, as nothing can be done to reduce the size of the dog population.

We have identified access to food and exposure to disease as two areas in which there is no reason to be complacent. The two are to some extent inter-dependent, since the consequences of disease are more serious for the hungry and the chronically sick are less able to produce food. Neither hunger nor disease can be eradicated by simply spending or handing out money; both present problems which demand knowledge, understanding, sympathy, and patience. Perhaps the most hopeful and realistic move to combat disease is the involvement of local people in the running of health clinics. In Tanzania, for example, where only a tiny fraction of the population has ready access to doctors and modern hospitals, young people have been given just sufficient training to identify and treat common ailments and to recognize and refer to doctors anything beyond their own capabilities. With such a system the few highly-qualified and expensively-trained doctors are more efficiently deployed than in a rich country, because they see only cases that demand their expert knowledge.

## The ecology of motor cars

Let us look again at the European consumer, and in particular at the motor car, to examine the widely held belief that everyone has the right to own and drive one. We shall wander a little from what is strictly understood as ecology, but no matter. Ownership of a car immediately places a person in a different position as a consumer of raw materials, and is perhaps one of the most obvious differences between the average European and the average African. The car alone probably contains more raw materials than the total amount a person in a poor country can expect to consume in a lifetime, and to this we must of course add the consumption of petrol and oil in quantities far in excess of what the average African can expect to have available.

Most of us live in or near towns and cities and we are familiar with the problems created by the urban environment. We are all aware that the motor car is both useful and a menace. Pollution, it has been facetiously suggested, is caused by the noise and the fumes of other people's cars, not our own. But most of us have been persuaded that the solution to the motor car nuisance lies in building better roads and motorways.

The manufacture of motor cars is an important business in industrial countries and there is intense competition for overseas sales. Cars are a major export, and therefore important to trade for several countries, notably Britain, France, Germany, Italy, Sweden, the United States, and Japan. As is well known, there are many different makes of car and we are persuaded by elaborate and expensive advertising campaigns to buy this or that model and to be always on the look out for a better car than the one we already have. Cars are not built to last – they are intended to be replaced within relatively few years of purchase – and the entire industry and sales campaign is geared to rapid production and consumption, and, it may be added, to rapid decomposition, because once a car is abandoned as no longer functional rather few of its materials are recycled to manufacture new cars. Making cars is thus extremely demanding on the supply of raw materials, especially metals, oil, and rubber.

There are now about 18 million motor vehicles in Britain, or one to every 60–70 metres of road. About 1·7 million new cars are produced a year and 600,000 are scrapped; many of the new cars are exported, but imports of foreign cars roughly balance exports. Since 1973 the cost of cars have trebled while average incomes have not. Despite a fall in sales in the few years after the 1973 oil crisis the car population of Britain is increasingly rapidly. At the present rate of increase there could be 40 million cars in thirty years time, an impossible figure that does not, of course, take into account the fact that raw materials for manufacture and operation are unlikely to be available for more than another twenty years or so. The increase in road traffic has led to demands for more and better roads. Already about a quarter of the area of London consists of roads, and this fraction is increasing. The cost of roads, especially of motorways, is astronomical, as also is the quantity of materials used. A motorway interchange requires 26,000 tons of steel and 250,000 tons of concrete, and substantial sources of energy in the form of oil to move these materials and to structure the landscape.

In the European Common Market countries about 60,000 people are killed every year in road accidents. It is estimated that nearly half the children in western Europe can expect to be involved in a road accident at some stage in their life. A disease causing deaths and suffering on this scale would be subject to massive attempts at eradication, but the car increasingly dominates the scene and every effort is made to increase its numbers and its special environmental requirements. The car is a voracious consumer of non-renewable resources, and a hazard to our lives, and yet through economic circumstances no longer possible to control, it is an essential part of industrial man's way of life.

Few people would disagree that motor cars are a major source of pollution and of noise especially affecting people living in cities or near motorways. Millions of tons of carbon monoxide are emitted from motor traffic every year as well as substantial quantities of oxides of nitrogen, sulphur dioxide, ethyl lead, vanadium, rubber and asbestos dust, and hydrocarbons. About forty per cent of the lead produced in the world is used as a petrol additive. Lead is one of the nastier potential sources of poisoning of the environment and of people.

All things considered, the motor car is one of the most important sources of pollution and is for instance a far greater threat to health than the smoking of tobacco. And yet little is being done to slow down the rate of production of cars and to restrict their use; on the contrary more roads are being advocated and the ups and downs of the car industry (which in Britain is regularly brought to a standstill by strikes for more wages) are a major political and economic issue in industrial countries. From what we know of density-dependent factors regulating natural populations of organisms, it should be clear that the solution to the car problem does not lie in building more roads as this will simply enable the car population to increase even more rapidly. In Britain the construction of the first motorways in the late 1950s was hailed as a major technological break-through, but already the motorways are congested, which has led to demands for more motorway construction. There is no obvious end except that we can be sure that whatever steps are taken now to support the road traffic industry they will not be sustained, because of the car's complete dependence on oil. In Britain alone about 13 million tons of petrol are consumed every year and, of course, none is recoverable.

The world's supply of oil is derived from sedimentary deposits of fossil animals laid down many millions of years ago. These deposits occur in various parts of the world and, somewhat ironically so far as the industrial nations are concerned, the best deposits are in under-developed non-industrial nations, especially in the Middle East. This accident of nature has stimulated the formation of gigantic national and multi-national oil companies which prospect and produce oil and market oil products, including petrol. Some small countries like Libya with good oil reserves have become immensely rich, and since they have a monopoly on the supply they continually exert pressures to raise the price of crude oil. It has been suggested that some oil should be left unexploited and in reserve on the grounds that it will appreciate in value and is a better investment for the future than anything else, but the general trend is to get the oil out and sell it as quickly as possible. It may be noted that some of the countries with large oil reserves have not thus far spent significant amounts of their profits on social reforms: many politicians, businessmen, kings, and other royalty have become immensely rich, but they have shown some reluctance to distribute wealth to the bulk of the people in their countries, or to the under-developed countries that lack oil. Oil from the North Sea will provide for most of Britain's requirements in the next twenty years, and will maintain the nation's economic position, postponing the need to look seriously for alternative sources of energy.

The demand for oil is increasing exponentially. Much of the industry in developed counties is dependent on it, yet there is clearly a limit to how much can be extracted. Estimates vary, but it seems certain that there will be a marked shortage of oil by the end of the century and that this will be preceded by rising costs, perhaps to such an extent that individual ownership of motor cars will become impossible within a few years except for the very rich. The big oil companies are powerful and a considerable amount of world power is intricately bound up between these companies and the governments that support them. The oil companies are interested in profits and we must assume that they are aware that what is happening now cannot be expected to go on for much longer. The demand for oil is likely to exceed supply in a matter of twenty or thirty years and the scramble to gain control of what is left is intensifying, not only in terms of economics but also politically. It may only be a matter of time before a major war is initiated over oil supplies.

The motor car's prospects are therefore precarious. Nothing but instability can be predicted for its future. Cars are at the moment increasing in number and density in all industrial nations, and space is being created to accommodate them by destroying land and building roads. They require a source of energy that will soon become unavailable, and in the end they will disappear, perhaps as quickly as they appeared. In the meantime they will continue to provide cheap transport, and will continue to pollute and kill. We shall insist until the last moment that it is our right to own and operate a car and we shall evade alternatives – after all, twenty years is a long way ahead.

## Rooks and squirrels

Our study of the ecology of the motor car suggests that its impact on society will fall off rapidly as its source of energy becomes expensive and unavailable. We could equally well proceed and examine the aircraft industry and indeed other human enterprises that involve the use of non-renewable resources, but we shall instead change the subject and discuss our relationships with animals widely regarded as pests, which, despite enormous efforts, have shown themselves able to resist our attempts at destroying them.

Many animals fit into this category and all of them share the ability to exploit ecosystems created by man. The two we shall discuss, the rook and the grey squirrel, are in Britain considered worthy of the expenditure of much time and money in order to destroy them and to try and control their numbers. The rook is regarded as a threat to agriculture because of the amounts of grain it is reputed to eat and the vegetable crops it destroys, the grey squirrel is a threat to forestry because of the damage it does to growing trees.

Rooks nest in colonies in tall trees and their calls and activities are a familiar feature of rural areas. They are almost entirely dependent on agriculture and must have increased enormously in numbers after the ancient oak forests were cut down and the land cultivated in the Middle Ages. Indeed it is likely that rooks were rare in Britain before the forests were destroyed. They occupy traditional nesting sites year after year, and in early spring they assemble to repair and build nests in exactly the same spot as in previous years. The clutch-size varies between two and seven eggs, most individuals lay either four or five, and egg-laying usually starts in early March. Breeding occurs early in warm and late in cold springs, the maximum difference between years being about two weeks. Early breeding ensures that the young rooks are being fed by their parents at the best time of year for the kind of food they require. In most years food is readily available on pasture (provided there is some rain), the adult birds collecting for their young invertebrates like earthworms, the larvae of daddy-longlegs flies, some grain left over from the previous year's harvest, and some newly sown grain. If there is an exceptional drought, many of the invertebrates become unavailable, the earthworms in particular retreating deep into the soil where they cannot easily be obtained. If it is exceptionally warm and wet, the vegetation may grow up quickly, making feeding difficult because rooks need short grass for optimal feeding and do very badly in rank vegetation. Hence nesting success defined in terms of young produced per nest varies somewhat between years: it is high in springs that are rather wet but not too warm and low if the weather is

*Plate 23:* A rook's nest.

dry and cold. Up to a third of the young that hatch die of starvation in bad years, while in good years nesting success may be much higher, although broods of five or six seem almost invariably to lose one or two young from starvation. The rook's nesting success is therefore re-miniscent of that of the heron described in Chapter 3.

When they are fully-grown the young rooks leave their nests and move out onto the branches of the trees, although their parents continue to feed them. It is at this juncture that farmers indulge in a highly ritualized pastime. With their friends they visit the colonies and with shot guns and rifles pick off the young rooks from the branches one by one, often killing every young bird produced in the colony. The justification for this slaughter is that in this way the rook's numbers are controlled, but the sporting element is strong, and until recently rooks provided the essential ingredient for a late spring feast: they are after all the four and twenty blackbirds baked in a pie in the nursery rhyme 'Sing a song of sixpence'. Not all colonies are visited by farmers, and considerable numbers of young birds leave successfully.

Those young rooks that survive the shooting now face the most difficult time of their lives. The rook feeds where the vegetation is low or where the fields have been ploughed, but by the summer such places are scarce. If the weather is dry the earthworms are deep in the soil, caterpillars have turned into moths, and the rooks face a serious food crisis. Many of them, perhaps more than half, die during these first few months. As winter approaches the fields once more become accessible to feeding birds as many crops have been harvested, the vegetation is short again, and the weather tends to be wetter. Provided there are no severe frosts to reduce the availability of animals near the surface of the soil, winter is not a bad season for rooks. Early in the winter large numbers of migratory rooks arrive from continental Europe and swell the resident population. The resident birds disperse about the countryside, some moving a long way from their breeding colony, and British and continental birds join together to feed in the fields and to roost communally. During winter a few rooks are shot by farmers and some by game-keepers who mistake them for crows, but the main effort at destroying them is confined to the late spring when the young can be easily slaughtered.

It is widely believed that the annual rook shoot is necessary, and millions of young birds have been destroyed in this way, at considerable expense. Despite this onslaught the rooks until recently maintained and even increased their numbers year after year and there was no evidence whatsoever that shooting had the slightest effect. Why was this? The answer is really quite simple: although many birds were destroyed, the destruction merely anticipated the high death rates that occur in the summer, the lean season for rooks. Possibly, if young rooks were not shot, the summer would be even worse, as there would be more competition for the available food and perhaps proportionately higher death rates. It thus seems possible that shooting the young birds at the colonies keeps the population at a higher level, as it decreases the natural death toll that occurs each summer. For shooting to be effective it would probably be necessary to destroy about ninety per cent of all the young rooks produced in Britain each year. This is obviously an impossible task, and even if it were possible it is likely that the countryside would be colonized by immigrants from Europe. Britain's rook population has recently declined and today is more than forty per cent lower than it was in the mid-1960s. The decline, especially in the last few years, is correlated with the dramatic reduction in the number of elms which rooks favour as nesting sites. It is also correlated with an increase in the amount of land under cereal cultivation and a decrease in the amount of pasture. We do not know if either or both of these events have been the direct cause of the rook's decrease; if the cause is the increase in the amount of land devoted to cereal cultivation, then the rook can never have been a serious pest of grain crops, since if it were then its numbers ought to have increased.

The rook is so well-adjusted to man's environment that it is now probably impossible to dislodge it, and in any case many country people would regret its disappearance. The population dynamics of the rook are currently being investigated; thus far the results seem to confirm what is already known, and it could be seriously doubted whether research of this sort will ever be of economic value, although this is not to question the intrinsic interest of the research as a piece of scientific investigation. We still do not even know for certain why the rook has recently declined in numbers. There is, however, much scope for research into the folk-lore associated with rooks and into the beliefs of farmers and country people who really seem to feel that the annual shoot is necessary to prevent widespread destruction of their crops. But no matter what is discovered, the rook is now an integral part of the countryside, and perhaps like its plant counterpart, the buttercup, its presumed bad points can be ignored and its attractiveness appreciated.

The rook's success can be viewed in terms of its intimate association with farming and with the alteration of the landscape following the destruction of the forests hundreds of years ago. Our second species, the grey squirrel, has a different history. There is no record of the grey squirrel in Britain before 1876 and it was not until 1890 when ten were released in Woburn Park, Bedfordshire, that it became established. Reputedly this particular introduction was tried because the local gentry had run out of things to shoot, but whatever the reason it is clear that the grey squirrel, a native of North America, has established itself.

*Plate 24:*  Rooks' nests in a dead elm. The recent decline in the number of rooks is correlated with the loss of trees killed by Dutch elm disease.

No one could have predicted ninety years ago that this attractive little animal would cause so much alarm to forestry interests in Britain. At least thirty successful introductions were made during the thirty-five years following 1890, and what had at the time seemed a harmless experiment generated one of the biggest population explosions known for an animal of this size.

The grey squirrel has now spread all over Britain and wherever it has become well established, the only indigenous species, the red squirrel, has declined in numbers and usually disappeared. This suggests that the two species compete and that the red squirrel is a consistent loser, but it is by no means certain exactly what they compete over except that at certain times of the year when food is scarce the grey is better at exploiting what is available than the red. Curiously, although British foresters condemn the grey squirrel as a serious pest of trees, it has not earned this reputation in its native home in North America. Grey squirrels attack the bark and thus inhibit growth and sometimes cause the death of a tree. They also damage twigs and branches, remove tree seeds that have been planted in forestry nurseries, and take the eggs and young of wild birds. They are further condemned because they are aliens, and because they are said to bully the red squirrel, two attributes that are not well tolerated by British naturalists and countrymen. As long ago as 1931 the National Anti-Grey Squirrel Campaign was set up and the destruction of squirrels was strongly advocated by the government. So-called 'squirrel clubs' were organized and cartridges were issued free in order to accelerate the extermination of the alien. But the squirrel had already become abundant and even though hundreds of thousands were shot the species not only held its own but continued to increase. In 1952 nearly 170,000 grey squirrels were killed in England and Wales with no noticeable effect on overall numbers. It now occurs through-out the country and is as familiar in towns as it is in rural areas, having invaded city parks and gardens where it has exploited the natural generosity of residents who feed it with bread. In 1973 it became legal to poison grey squirrels and it will be interesting to see if a programme of poisoning has any significant effect on its numbers; we can certainly expect an outcry that other animals and perhaps people will be poisoned accidentally if this campaign is developed.

The dilemma of what to do about the grey squirrel has many similarities with that surrounding the rook, except that unlike the rook

*Plate 25:*  The grey squirrel.

it was a deliberate introduction. Enormous numbers of squirrels and rooks are shot but with no effect upon overall numbers; both species are now part of the British countryside and although it may be true that both cause damage it is probably cheaper and perhaps more efficient to leave them alone. In any case, we may nurture a feeling of admiration for animals like rooks, squirrels, and locusts which have successfully resisted our technology and scientific knowledge and which flourish despite our attempts to exterminate them.

## What can be done?

Although man must be considered as a temporary occupant of the world ecosystem, his impact has been far greater than that of any other species since life began. We have successfully converted much of the world to suit our own purposes, and yet at the same time we are constantly frustrated by other species; the rook and the grey squirrel are merely two examples of a common phenomenon. The short answer to the question of what can be done is that we must at least think carefully about the long-term consequences of population growth and of a consumer-oriented society which generates pollution and uses up the world's non-renewable resources. If you think in ecological terms the answer is clear: all exponential processes must eventually stabilize and the only remaining questions are how stability will be achieved, and, of course, when.

One of the dangers of the present world scramble for resources is that prices for raw materials may rise so steeply that it will soon cost more to import them than we can expect to earn by exporting the finished product made from them; this in itself should result in a slowing down in the rate of economic growth. We have in this chapter identified the dwindling supply of energy, especially in the form of oil, as the most likely event that will force the industrial countries to limit their economic growth rates, and there is every reason to suppose that this will begin to have effects within the next five or ten years. In contrast, it is by no means clear what will happen to world population growth except that we can expect rates of increase to slow down in industrial countries and to go up in poor countries, events which are likely to have widespread political and economic implications. India will probably be the first nation to give us an insight of what may become a world phenomenon, and as we have discussed earlier, it is probably only a short time before we can expect to hear of enormous death rates from starvation and disease.

Thus it would apear that industrial man's role as an exploiter of the rest of the world will shortly receive a severe jolt, triggered initially by a failure of oil supplies, and followed by the consequences of over-population in poor countries. But perhaps by placing man in an ecological setting we shall at least understand our predicament more clearly. Unfortunately, many people in the world are either unable or unwilling to face up to the problems.

People in the rich countries must learn to recycle discarded materials and to avoid waste; indeed the idea of waste springs from the perception of most by-products of human activity as useless. Non-returnable bottles discarded without thought by Europeans would be carefully used and re-used by African villagers. Only now are we beginning to learn just how wasteful our society has become; for instance from 1958 to 1970 per capita milk consumption in the United States decreased by 23 per cent but during the same period the amount of material used for milk containers increased by 26 per cent. There is absolutely no excuse for this profligacy, yet Britain, which is one of the few countries to market milk in returnable bottles, is being urged to adopt throw-away milk containers in the interests of harmonization of Common Market policies. Estimates vary and there are differences between countries, but something like 70 per cent of the metals extracted are used but once and then discarded. Apart from making better use of non-renewable resources, recycling of scrap iron and steel by mills and foundries leads to an 80 per cent reduction in air pollution, a 76 per cent reduction in water pollution, a 40 per cent reduction in water use, and effectively eliminates the production of solid wastes. In nature, as we have seen in Chapter 5, everything produced is eventually used; waste is a human concept, and we are the only organism capable of making use of our waste manufactured items. There are encouraging signs in some rich countries that the problem of waste is understood and there are moves towards a more sustainable society based upon recycling, re-use, and repair.

Each nation is justifiably proud of its technological breakthroughs. Many of us admire the appearance and performance of Concorde but in social terms the project is a failure: it is a fine technological achievement but a financial disaster, benefiting only the very rich at the expense of the ordinary tax-payer. But the ordinary tax-payer can benefit from technologically less spectacular but socially advanced projects. In the 1950s it was realized that the River Thames flowing through London was so polluted that it was biologically dead; it was nothing more than a gigantic sewer carrying industrial and domestic waste into the North Sea. The black, smelly water contained no oxygen but plenty of poisonous chemicals. In the 1960s improvements were made to sewage works and new laws were introduced to restrict the discharge of industrial waste into the River. As a result the level of oxygen in the water rose rapidly and fish began to be found in places where they had long been absent. By 1976 about ninety species of fish had been recorded in a stretch of the River that in 1957 had supported only the eel. Seaweeds have established themselves in the tidal waters and large numbers of duck have returned to feed in the much cleaner water. There are even reports of salmon being sighted, and there are plans to introduce young salmon into the upper reaches in the hope that the Thames will once again provide a breeding site. Perhaps the Thames may soon return to the bountiful state it enjoyed in the Middle Ages when London apprentices were protected by their indentures from having to eat Thames salmon more than twice a week.

The transformation of the Thames is a story of successful action. Londoners are no longer appalled by the stench of the River and there is no reason why the successful clean-up of London's waterway should not be copied elsewhere. What to do and how to do it are known; all that remains is the will to turn ideas into action.

The aim of this book has been to explain the principles of ecology in ordinary language in such a way that some insight into man's present dilemma may be obtained. It was not intended to provide answers and suggestions for action. That would require another book, more political in scope, but what we have discussed should provide a useful background, or at least something to think about. Action on environmental issues should begin at home, and perhaps the best approach is to ask questions of those people who advocate building more roads, cutting down trees, shooting seals, and destroying historic buildings in the interests of progress and profits. A beginning has already been made, and we may see the day when a new brand of politics is evolved which pays less attention to profits and losses and more to the quality of life.

# A guide to further reading

There are numerous books on ecology. Those mentioned below are a small selection of what is available, which should prove useful for anyone requiring additional information on the topics discussed in this book. All contain references to other works which take the reader deeper into the subject or enable specialization in one part of it.

I have not said much about how ecological information is acquired. As with all branches of science, information is obtained by observation, experiment, the analysis of results (frequently by the use of statistical methods), and by formulating generalizations and predictions which in turn lead to further observations and experiments. Two books, T. Lewis and L. R. Taylor, *Introduction to experimental ecology* (Academic Press, 1967) and T. R. E. Southwood, *Ecological methods* (Methuen, 1978), intended primarily for use in schools and colleges, explain how to find out how many earthworms there are in the lawn, how to sample insect populations, and many other things, and are invaluable to anyone wishing to undertake ecological investigations.

There are several large textbooks which contain a wealth of theory, backed with numerous examples from around the world. Two of the better ones are E. P. Odum, *Fundamentals of ecology* (Saunders, 1971) and R. E. Ricklefs, *Ecology* (Nelson, 1973). The characteristics of exponential processes are discussed in most books concerned with ecology, but are perhaps most clearly formulated in D. H. Meadows and others, *The limits to growth* (Earth Island, 1972). This book is regarded as controversial because it deals with exponential processes as they apply to man, including population increase, the consumption of resources, and the effects of pollution. It is a book for all thoughtful people, and although some of the assumptions on which the conclusions are based may be questioned, there is no doubt that it is of vital importance to an understanding of man's environmental dilemma. A more extensive treatment of the same subject is given in P. R. Ehrlich and others, *Ecoscience: population, resources, environment* (Freeman, 1977).

The structure of communities and ecosystems is discussed in most ecology textbooks; particularly recommended are R. H. Whittaker, *Communities and ecosystems* (Macmillan, 1977), which takes most of its examples from North America, and C. S. Elton, *The pattern of animal communities* (Methuen, 1966), with most examples from Britain, and with a detailed account of how to conduct an ecological survey.

The special ecological problems of under-developed countries are discussed in R. F. Dasmann and others, *Ecological principles for economic development* (Wiley, 1973) and D. F. Owen, *Man's environmental predicament: an introduction to human ecology in tropical Africa* (Oxford, 1973).

Finally, there is the problem of how species of plants and animals can be identified. There are numerous field guides and keys for the flora and fauna of northern Europe. Five books published by Hodder and Stoughton in 'The natural history of Britain and northern Europe' series provide a good start. Each contains an ecological essay about a specific habitat, an illustrated field guide, and a list of references to publications on groups of plants and animals. The books are: A. Darlington, *Mountains and moorlands* (1978), D. Owen, *Towns and gardens* (1978), D. Boatman, *Fields and lowlands* (1979), R. Barnes, *Coasts and estuaries* (1979), and B. Whitton, *Rivers, lakes and marshes* (1980).

# Index

*Other Paperbacks from Oxford*

## Man's Environmental Predicament
An Introduction to Human Ecology in Africa

D. F. Owen

'With the possible exception of man himself, insects are by far the biggest overall threat to economic development in tropical Africa.'

In the Western mind ecological problems have come to be associated with the now-apparent effects of industrialization – pollution, over-consumption, over-population, destruction or erosion of the countryside. It is, however, to a large extent biological factors that have continually handicapped the less developed areas of the world. Dr. Owen examines these factors in tropical Africa and the way disease, pests, and nutritional deficiencies are being combated to improve health, productivity, and economic well-being there. Unfortunately the results of the struggle are often counter-productive, leading to larger populations less naturally immune to disease and more susceptible to food shortage. Attempts to adapt diets or to instil different habits of hygiene meet strong cultural and religious barriers. Intensive cultivation particularly of single crops can provide ideal feeding and breeding grounds for pests. Meanwhile serious economic development is further inhibited by Africa's role as provider of raw materials to the developed countries. Most African nations are uneasily dependent for their income on the export of one or two commodities.

Dr. Owen's study of this predicament should interest agriculturalists, development planners, anthropologists, sociologists, and almost anyone concerned with Africa.

## The Facts of Food

Arnold E. Bender

Why do we have to eat? What are vitamins, amino acids, proteins? Where do we get them and what good do they do us? How were they discovered, and how did people survive beforehand? Can man live on fish and chips alone? Do the Russians make artificial caviare? These are just a few of the questions answered in this book, which will also tell you about what we need to eat and how much, about fashions and taboos, about natural foods and chemical additives, about obesity and heart disease, about food labelling and the law, and about the contribution that the science of nutrition can make towards easing the problems of feeding the world's growing population.

# Patterns of Sexuality and Reproduction

## Alan Parkes

Can we choose the sex of our children? Why are more boys born than girls? Does the quality of intercourse affect the chance of having a baby? Human reproduction is cumbersome, prolonged, and rather messy. It is no easy task to unravel the intricate patterns of the effects of cyclic, seasonal, and social variations on our sexual activity and our ability to reproduce. This unique account of these patterns, with its clear but entertaining style, will appeal to all, from the layman, whose enthusiasm is mainly practical, to the medical student and the General Practitioner, for whom the subject is of professional interest.

# Man Against Disease
Preventive Medicine

Muir Gray

'About 50,000 people still die each year in Britain from cigarette smoking, principally because of lung cancer, heart disease, and bronchitis.'

'Alcohol is a valuable, socially acceptable drug, but its side-effects are serious and have to be controlled.'

'Britain is currently faced with the problem of the disposal of nuclear waste which will emit alpha particles for 10,000 years.'

Do governments spend enough time and money in the most efficient manner on the prevention of sickness? Does the individual devote enough attention to his own good health? Dr. Muir Gray considers the scope for preventing disease and premature death in both developed and underdeveloped countries, from the promotion of child health to the problems of old age. Now there is a growing awareness that curative medicine cannot solve all problems, and if effective high-technology medicine is to be offered to as many people as possible, it is necessary to prevent those disorders which are preventable.

The ways and means of prevention, education, and legislation are discussed, and particular attention is paid to the obstacles to prevention: ethical problems of what degree of state intervention is acceptable; psychological resistance to unpalatable truths; linguistic and cultural barriers; and financial considerations of how available money should be spent. The author also considers hazards to health posed by changes in the environment, pollution, and stress, and suggests that to tackle them man not only has to modify his physical environment but also the society in which he lives. The quality of life has become an important political issue.

## The Philosophies of Science

### R. Harré

In this book the author shows how various views about the nature of science are related to the great historical schools of philosophy. This is not just an abstract exercise in analysis. The argument is set out in terms of certain concrete episodes in the history of science – a manner of exposition which brings out most clearly the influence of philosophical theories on the development of science, and of scientific discovery on modes of thinking in philosophy. The book is intended for both philosophers and scientists, and grew out of the author's introductory lectures to interfaculty audiences in Oxford and elsewhere. An unusual feature of the book is the prominence given to the place of metaphysics in the philosophy of science. Metaphysics is seen as providing a series of options among ways of thinking; and the choice of option determines to a large extent the kind of philosophy of science that emerges.

'R. Harré's *The Philosophies of Science* offers a respectably cool, hard look at scientific thought and its relationship with the great historical schools of philosophy . . . both scholarly and lucid . . . and as good an introduction to the subject as could be wished for.'
*The Times Literary Supplement*

# A Historical Introduction to the Philosophy of Science
*Second edition*

## J. P. Losee

This is a straightforward exposition of the views that have been held by scientists and philosophers, from Aristotle to the present day, on such subjects as scientific method, the nature of scientific laws, the relation of experiment to theory, the function of hypotheses, and the evaluation of competing theories. The book aims to make the philosophy of science accessible to readers who have only a small knowledge of formal logic and the history of science. This new edition includes an expanded presentation of work in the field since the Second World War.

## A Short History of Scientific Ideas to 1900

Charles Singer

This book, demanding a minimum of preliminary knowledge, presents in simple form the development of the concept of a material world, all parts of which are rationally interrelated. In placing the basic scientific ideas in a framework of world history, from the earliest times in Mesopotamia and Egypt until AD 1900, it treats not only the physical and chemical but also the biological disciplines. The book has grown from *A Short History of Science*, which was published in 1941 and rapidly accepted as a standard work.

'One reason why this new history of science is assured of an illustrious career is that it is a work of such consummate art . . . masterly in conception and execution.' *The New Scientist*

'. . . this book is in the very front rank. It seems less and less likely that any other such general work on the history of the sciences will be written by such a master of the art as Dr. Singer.' *Advancement of Science*

'Dr. Singer deserves well of Western man . . .' *The Economist*

# The Structure of the Universe

Jayant Narlikar

Do black holes really exist? Why does the Sun emit fewer neutrinos than expected? What is a quasar really like? Where do galactic radio sources get their energy from? Do the laws of physics, as we know them on Earth, really apply on an astronomical scale? Professor Narlikar looks at the Universe and its mysteries from two contrasting points of view. He describes how the modern astronomer investigates the structure of the Universe and interprets what he sees and he shows how the physical environment here on Earth is so totally dependent on the structure of the Universe that the study of that structure can provide valuable information for the earth-bound scientist.

'The name of Jayant Narlikar has long been associated with the steady-state theory of the Universe, the absorber theory of radiation and other aspects of modern cosmology. Now he has used his fertile imagination to produce a readable and useful introductory book to astronomy and cosmology . . . this is a nicely written, attractive book, including many useful . . . discussions of basic modern astrophysics, from an author of international repute.' Paul Davies, *Nature*

## The Solar System

Zdeněk Kopal

In the last decade radioastronomy, spacecraft, and
astronauts have revolutionized our knowledge of our
sister planets and their structures. The author
describes the impact of this new knowledge, in non-
technical language, with an authoritative account of
each of the planets, including the Earth, and the asteroids,
comets, and meteors of interplanetary space. He
discusses current views on the origin of the solar system,
the evidence for other planetary systems, and the future
of planetary research.

**Mushrooms and Toadstools**
A Field Guide

Geoffrey Kibby

This practical field guide gives accurate full-colour
illustrations of more than 400 of the larger fungi of
Britain and Europe, some of which, not all of them rare,
are shown for the first time in colour. Geoffrey Kibby
and Sean Milne collaborated closely in the preparation
of the paintings, which show each species in its typical
habitat and draw out its characteristic features. The
detailed descriptions on the facing pages are designed
for easy reference: the distinctive characteristics are
highlighted by bold type, and followed by a full account
of habitat, season and frequency of occurrence, and
details of the fruit-body and spores, as well as simple
chemical tests, and a note on edibility.

The introduction to the book acquaints the reader
with the world of fungi from botanical, taxonomic, and
economic points of view, and contains sections on
collecting and identification as well as on poisonous
species and the use of fungi as food. There is also a
uniquely illustrated key to lead the user to the appropriate
place in the book. *Mushrooms and Toadstools: A Field
Guide* is a valuable new book that no keen naturalist can
afford to be without.